The Boundaries of the New Frontier

Studies in Rhetoric/Communication
Thomas W. Benson, Series Editor

The Boundaries
of the New Frontier

Rhetoric and Communication at
Fermi National Accelerator Laboratory

Joanna S. Ploeger

The University of South Carolina Press

© 2009 University of South Carolina

Published by the University of South Carolina Press
Columbia, South Carolina 29208

www.sc.edu/uscpress

Manufactured in the United States of America

18 17 16 15 14 13 12 11 10 09   10 9 8 7 6 5 4 3 2 1

Library of Congress Cataloging-in-Publication Data

Ploeger, Joanna S.
  The boundaries of the new frontier : rhetoric and communication at Fermi National Accelerator Laboratory / Joanna S. Ploeger.
    p. cm.—(Studies in rhetoric/communication)
  Includes bibliographical references and index.
  ISBN 978-1-57003-808-2 (cloth : alk. paper)
  1. Communication in science—Illinois—Batavia. 2. Research—Social aspects. 3. Fermi National Accelerator Laboratory—Public relations. I. Title.
  Q223.P56 2009
  507.2—dc22
                                                                              2008045823

This book was printed on Glatfelter Natures, a recycled paper with 30 percent postconsumer waste content.

For my son,
Thomas William

Contents

List of Illustrations *ix*

Series Editor's Preface *xi*

A Note on Joanna S. Ploeger *xiii*
    David Depew

1 Interactions *1*
2 Mapping the Boundaries of the Laboratory *32*
3 Rhetoric, Persuasion, and the Sublime Laboratory *54*
4 Practice and Perception *83*
5 Stakeholders, Self-Tours, and Communication after the SSC *117*
6 Mapping the Boundaries and Charting a Future *152*

Notes *167*

Bibliography *185*

Index *193*

About the Author *201*

Illustrations

Fig. 1. "Caldron." Drawing by Angela Gonzalez 7
Fig. 2. "Broken Symmetry" 69
Fig. 3. "The Obelisk" 71
Fig. 4. "Tractricious" 72
Fig. 5. "Wilson Hall—the Power of the Sun." Drawing by Angela Gonzalez 74
Fig. 6. "Wilson Hall—Moonlight over the Prairie." Drawing by Angela Gonzalez 75
Fig. 7. "Wilson Hall and Obelisk Reflected in Pond." Drawing by Angela Gonzalez 76
Fig. 8. "Science at Fermilab" 126
Fig. 9. "Collision" 128
Fig. 10. "Is It a Top Quark?" 129
Fig. 11. A superconducting magnet 130

Series Editor's Preface

Joanna Ploeger's *The Boundaries of the New Frontier: Rhetoric and Communication at Fermi National Accelerator Laboratory* presents a rhetorical account of the way scientists at the lab communicate with one another and with a variety of publics. Fermilab is a government-supported high-energy physics laboratory in Batavia, Illinois, some forty miles west of Chicago, conducting advanced research, according to its self-description, in the "science of matter, space, and time." The lab also, in Ploeger's account, engages in an ongoing rhetorical exercise in the technological sublime, celebrating the achievements of science and offering public education in the work of the lab, bringing together two groups that have differing views of one another and partly shaping the lab's decision making. Ploeger writes that, "From the time of its founding in 1968 to its prominence following the discovery of the top quark in 1995 and on to the present day with questions about a future role in particle physics research, Fermilab's evolution as a research institution has been strongly influenced by competing cultural narratives that attempt to define what it means to do science in the public interest." Ploeger's book traces the rhetoric of those cultural narratives among the scientists at the lab and its various publics, through observation, interviews, and analysis of a wide range of contemporary and historical cultural texts, written, spoken, experiential, and visual.

Ploeger's interviews with scientists reveal how boundaries of identity are maintained and negotiated and how individual scientific commitment competes with the increasingly complex and bureaucratic matrix in which large-scale physics is conducted. Physics at this level requires vast amounts of public money to support research in which physicists work in large teams, subordinating the work of the individual scientist. Constant attention must be given to the funding of these efforts.

For their part, members of the nonscientific publics who visit the lab have their own complex views of the work done there. Ploeger describes the education and public-relations operations of the lab and studies both the exhibits and presentations on the lab's public tours and the responses of various visitors to those presentations. What she finds are some marked differences of perception across the boundaries between science and its publics. Visitors were more interested in the practical benefits that might emerge through new technologies enabled by science at the lab. The lab and the scientists, on the other hand, continued to present themselves as part of an exercise in scientific utopia and the technological sublime. Ploeger concludes that, "If public discourse about the future of science is to move forward, some common ground must be found between the sublime mythologies of the physics community and the increasingly powerful consumerist model that is evident in the comments of nonscientists."

Joanna S. Ploeger had nearly finished work on a revision of her manuscript for this book when she died of cancer in July 2006. Several friends and colleagues helped to bring the manuscript to completion; the text was edited by her former colleague, Professor David Depew of the University of Iowa.

Thomas W. Benson

A Note on Joanna S. Ploeger

A revised text of this book was nearing completion when its author was stricken with cancer. Joanna S. Ploeger died on 19 July 2006, in Berkeley, California.

The book is a detailed study of the communicative practices of the Fermi National Accelerator Laboratory in Batavia, Illinois—Fermilab, for short. The resulting study is unique for its assessment of the role of aesthetics in mediating the relationship between science and society at Fermilab and for its commendation of a new model of scientific communication, in which democratic publics are figured as more than sources of funding, objects of public-relations solicitations, and presumed opponents of scientific curiosity for its own sake.

Many hands went into preparing the manuscript for publication. Joanna had rewritten it in accord with suggestions by Barry Blose, her acquisitions editor at the University of South Carolina Press, and by her developmental editor in California, Fern Burch. Joanna's friend Andrew Bell located the final working copies of each chapter on her computer and secured the illustrations from her papers. Rhianna Wisniewski at Fermilab was most helpful in providing high-resolution files of the pictures and in assisting through the permission process. David Depew, her colleague throughout the years Joanna served as an assistant professor in the Department of Communication Studies at the University of Iowa, edited the text. He had considerable help from four graduate students: Alessandra Madella, Jason Moyer, Jane Munksgaard, and Peter Morningstar, who drafted the index. Funds for their efforts were arranged by Kristine Fitch, chair of the department, and by Iowa's interdisciplinary Project on the Rhetoric of Inquiry, on whose board of directors Joanna served.

The final manuscript is as faithful as possible to Joanna's revised text, although in the last chapter, especially, some guesswork based on her notes was required. Professors Frederick Skiff and Usha Mallik of the University of Iowa's Department of Physics and Astronomy offered help wording scientific issues. It should also be noted that Joanna's death foreclosed the possibility of tracking the most recent history of Fermilab and of high-energy physics in general.

Those of us who have worked to bring the book to press take ourselves to be erecting a memorial to a wonderful teacher, friend, and colleague. We are acting on behalf of all of Joanna's colleagues at Iowa and at California State University, Stanislaus, where she had begun the first year of a tenure-track appointment when illness struck her. We are also acting in the name of the larger community of her colleagues in the developing field of rhetoric of science and technology. Trained in this field by several of its pioneers, especially Tom Lessl and Celeste Condit of the University of Georgia, Joanna served as vice president and then president of the Association for the Rhetoric of Science and Technology (ARST). A memorial session on her work was held at ARST's annual meeting in 2006. Each year ARST offers a Joanna S. Ploeger Outstanding Essay Award.

Above all, the book is offered in the name of Joanna's students. Undergraduates and graduate students alike appreciated the clarity and vigor of her teaching style as well as her fierce support of their individual aspirations.

Joanna is survived by her mother, Betty; her sister, Amy; her husband, Robert Winovich; and their son, Thomas William, who was an infant at the time of his mother's death.

DAVID DEPEW
Project on the Rhetoric of Inquiry and
Communication Studies
University of Iowa

ONE

Interactions

If you asked a high-energy physicist what they study, they might say, "Particles: protons, electrons, quarks, leptons." More probably would they say, "Interactions." The function of modern particle accelerators and colliders is to orchestrate interactions among various types of particles and to observe the outcomes. As they sift through the myriad results of interactions produced with each particle collision, high-energy physicists discover clues to the most fundamental frameworks of the physical world.

This book is about interactions of another sort as well—communicative interactions among people who worked and visited Fermi National Accelerator Laboratory in Batavia, Illinois. Like interactions within particle accelerators, communicative interactions also produce exciting and often inexplicable results, and none more so than those shaped by the peculiar contours that the relationship of science and society takes within the space of the contemporary research laboratory.

Rhetoric in the Relationship between Science and Society

"No, I'm not a physicist." I must have uttered this sentence at least a hundred times during my first seven-month stint conducting research at Fermi National Accelerator Laboratory. When interviewing visitors to the laboratory, they would invariably ask, "Are you a physicist?" I would explain that I was a communication studies scholar interested in what nonscientists think about science. Visitors' reactions to this response were mixed. Some seemed relaxed and were even more eager to talk with me, relieved to know they would not be quizzed about their knowledge of physics. But others seemed almost disappointed, as if they had hoped to meet a "real scientist" during their visit.

Not surprisingly, the same question surfaced when I met with scientists and staff at the lab. Many inquired about my academic training and how I came to be so interested in physics and research laboratories. I would explain that I had always been fascinated by physics and that my own attempts to better understand the field had led me to make science communication a focus of my research. In these instances my interactions were defined as much by my relationship to the laboratory, the field of physics, and science as a whole as by my relationship to the individual sitting opposite me.

Over the course of my travels through the laboratory system, I traversed not only the boundaries between self and institution, self and other, and self and science, but also the boundaries that distinguish scientist from citizen, laboratory from community, art from science, and, of course, science from society. Whether with a physicist, technician, receptionist, or visitor, it was clear that interaction was somehow both transformative and definitive. At times my presence as an outsider within the lab disturbed some well-established categories and boundaries. As a communication researcher, I was accorded a status and freedom greater than that of the usual visitors, and yet I didn't fit neatly into any existing category on the organizational chart. Was I an outsider at all? After all, Fermilab is a federally funded facility, and I am a taxpayer. As I conducted my research, I often found myself awash in discourse intended to remind me this was "my" laboratory. Yet my greater access made me even more keenly aware of the dividing line between the world of the laboratory and world outside its gates. Even as the lab's discourse reflected the importance and difficulty of sharing research with various constituencies, it also constructed and closely guarded the identity and boundaries of the laboratory—its inner life.

It soon became clear that to understand Fermilab I had to confront its boundaries and spaces. Some were material, such as the boundaries of the laboratory grounds and the spaces those boundaries construct and contain. Others were more clearly discursive, such as the boundary between "research laboratories" and "weapons laboratories." Some seemed organizational and managerial, defining the contours of the laboratory's research agenda and technology. All these boundaries are rhetorically constructed. They mark and constitute the laboratory, its employees and its publics. They define the world of high-energy physics and help determine the outcomes of economic, political, and symbolic disputes about the laboratory and its mission. Most important, these rhetorical boundaries help to determine the power and importance of the national laboratories as institutions, justifying their work and negotiating their symbolic importance in debates about the role and reach of science in society.

Although this is a book about boundaries, spaces, and interactions, it is also a book about the people who populate the spaces on both sides of the laboratory's boundaries. I had the opportunity to interact with many of these people through interviews with lab employees and visitors. Some experiences were accessible only through extensive historical research or through the many stories that circulate in and around the laboratory. The lab has shaped the daily lives of many, particularly those scientists and staff members who have devoted much or all of their professional careers to it and its work. Its boundaries are manifest in the narratives these individuals tell about the laboratory. But the laboratory's influence is also felt outside its gates, from the communities adjacent to the Fermilab grounds to the halls of Washington. A legislator voting on the next federal science budget or a parent taking a science education program at the laboratory also have stories to tell about that laboratory. As I cataloged these many stories through the interviews I conducted and the documents I collected, it became apparent that there was no single or simple boundary that defined the space of the laboratory. Instead we find many discourses authored by a culture sometimes deeply conflicted about science. These narratives have shaped and reshaped the institution over the course of its forty-year history.

In sum this research fulfills two principal goals. First it provides a detailed description and analysis of the rhetorical situation surrounding modern high-physics research, one of the last bastions of "big science" in the United States. Second, by examining the rhetorics the community uses to protect its autonomy and authority and the public response to those rhetorics, this analysis enlarges our understanding of the rhetorical processes of boundary work and its relationship to other forms of scientific rhetoric.

Imagining the Laboratory

The rhetoric of Fermilab is dominated by image and narrative rather than argument. Rich in idealism and nostalgia, the laboratory's discourse often seems focused on the maintenance of its connection to the technological sublime, a concept that will become clearer as we proceed. The boundaries of the laboratory are defined through its unique articulation of architecture, open spaces, artwork, and techno-scientific myth. Some of the most evocative images of the laboratory can be found in the illustrations of the laboratory's resident artist, Angela Gonzalez. Brought to the laboratory by its first director, Robert Wilson, Gonzalez worked with Wilson to create a distinctive aesthetic for Fermilab. Offering an unusual mix of high-modernist illustrations and frontier mythology, some of her most striking contributions can be found among the hundreds of black-and-white illustrations that celebrate the people, places, technology, and achievements of the laboratory (fig. 1). One popular illustration depicts the laboratory as the incantations of a magician and his minions. Gathered around a caldron, they seem to have called forth the laboratory from the depths of imagination. The viewer may wonder about these characters or speculate about the spells or oblations that have been offered in support of their creative process. The fanciful nature of the creation, however, makes one thing clear—this laboratory is no mere instrument or research tool. The image does not celebrate practicality or parsimony, but rather references the magical, even mythical, qualities of the laboratory and its work.

Many of her other illustrations have since disappeared from public view at Fermilab. Gonzalez's creation images epitomize the mystery at the heart of our experience of the laboratory and in particular our cultural understanding of the federally funded system of national laboratories generally.[1] How is a public scientific institution such as Fermilab created and sustained through action and perception? Gonzalez's image strikes a chord precisely because there is indeed something magical and somewhat unbelievable about the fact that we, as a society, should choose to create and continue to support such an elaborate scientific and technological enterprise—a speculative creation with *no obvious practical purpose*.[2] Some might argue that Fermilab is the predictable outgrowth of American technoscience—just one more link in a chain of ever larger institutions. It is difficult to refute that characterization, and yet such deterministic accounts tell us little about why or how society builds its scientific institutions. These and other images work to emphasize the distinctions between the work of the laboratory and the mundane world outside its boundaries. Put simply, the images depict the laboratory as a space beyond the realm of everyday experience, a liminal world between the mystical cosmos and the merely mundane.

The boundaries constituted by these images are reinforced in the many narratives that give meaning and structure to life within the laboratory. In a speech marking Robert Wilson's eightieth birthday, Fermilab theorist Chris Quigg related a conversation with a colleague that seemed to echo the themes of fascination and mystery. Here Quigg pays homage to Wilson, undoubtedly among those stirring the pot:

> Toward the end of a very cerebral, yet seductively earthy, bottle of Zinfandel, I asked Peter [Peter Limon, Fermilab accelerator physicist] to tell me what he likes best about the Tevatron. Peter chewed his Zinfandel meaningfully. Then he said, "The neatest thing is that you can actually store up all those antiprotons day after day after day and put them in the machine when you want to.... You say, I'm going to put p-bars in the machine at midnight tonight, and you do that, and you accelerate them, and you squeeze the beams into little tiny spots at B0, and you bring them into collision. I'm not only fascinated—every time it happens; I'm surprised." Peter poured himself the last of the wine. "I think it's really spectacular," he said, "particularly since I know these people who are doing it." He rolled his eyes, then became very serious. "You know, if Fermi and Feynman and the god were doing it, you wouldn't think anything about it, right? But normal, everyday human

beings are doing this. What's amazing is that regular people get
together and make these amazing things work." To Bob Wilson on
his eightieth birthday, to all the builders and makers of this inspiring
place—this place of boundless horizons—thank-you for giving us
normal, everyday human beings the chance to discover, within
everyone of us, a little bit of the gods.[3]

This narrative does not celebrate the practical geniuses of the accelerator, but rather its mysteries and other-worldly nature. The physicist's relationship to the machine is one of communion rather than use. His identity is simultaneously distinct from, and yet defined by, the machine and its magical properties. Although U.S. physicists have long celebrated the greatness of their technological achievements, such ritual celebrations of the machine and its makers have a special significance at Fermilab. The art, architecture, and stories of this particular laboratory honor the accelerator as evidence of the creativity and vision of American science. Not every lab does that, or even tries it. But what are the consequences of the laboratory's construction of its "otherness?" Is the rhetoric productive or problematic? Such questions can only be answered through careful analysis of the images and narratives that mystify the character and purpose of the laboratory and, as a result, obscure the institution's roots in the history of physics and its sometimes-contested relationships with the publics it serves. The laboratory's consistent evocation of sublime narrative and imagery stands in sharp contrast to more traditional arguments offered in favor of basic research. This distinction raises questions about how the history of this laboratory and also the evolution of hadronic colliders have been informed by narratives that emphasize the mystery of our cultural relationship to technology. While the lab's very existence offers evidence that such rhetoric has proved persuasive in the past, this book calls into question the current and future efficacy of such a strategy, particularly in the face of changing historical circumstances and counter-rhetorics that interpret the laboratory's spaces in new ways and define its publics through and across different boundaries.

This study prompts reconsidering of many of our assumptions about the relationship between science and society. For too long, researchers have framed public participation in scientific decision making almost exclusively in terms of "public understanding," as though the only legitimate role for nonscientists was to understand and, by implication, appreciate the work of scientists. This study in no way rejects the importance of public understanding and concomitant efforts to improve both formal and informal modes of science education, but I hope that its findings will stretch our interpretation of the nature of public understanding and participation in science. First, the comments of study participants reveal the degree to which many in the sciences hold stereotypical and often inaccurate opinions about the public's interest level and capacity for intelligent deliberation about scientific issues and the fate of scientific institutions. Second, the study demonstrates the general public's willingness to educate themselves about scientific issues and institutions. The very fact that so many individuals voluntarily tour FNAL on a yearly basis suggests that nonscientists of many stripes find the facility and its work sufficiently important to warrant their attention and thoughtful consideration. Last, the comparison of the comments of these two groups, in conjunction with the historical analysis offered, suggests that the lab's decision-making processes are inevitably shaped by public discourse.

Imagining "the Public"

The chapters that follow focus not simply on rhetorics originating within the laboratory, but rather on the constitution of and communicative interaction among the lab and its publics. By studying discourse about Fermilab from the perspective of those both "inside" and "outside" the laboratory community, it is possible to understand better how publicly funded laboratories such as Fermilab are constituted and sustained through the rhetoric of scientists and nonscientists alike.

Some might question the choice of a high-energy physics laboratory as a site through which to explore the rhetorics that define the relationship between science and society. Fermilab was purposely built to reflect a constructive, positive, rather than destructive, application of physics research. Over the course of many years, however, the nature of public involvement in the laboratory system changed. As a result physics institutions of all varieties have seen their epistemic (or knowledge-producing) and social authority challenged. Early in its history, Fermilab attempted to make strategic use of public challenges in an attempt to secure a site and funds for the facilities that would determine the future of the field. Antiscience sentiment, growing strong at the time of the laboratory's founding, was met by decidedly humanistic rhetoric that contributed directly to creating new audiences for physics. In 1978 Robert Wilson authored *The Humanness of Physics,* which epitomizes the ways in which FNAL's rhetoric responded to problematic counter-narratives about physics. In doing so, it recasts the function of the laboratory and its relationship to the larger culture in a way meant to erase a negative image deposited by the militarization and commercialization of the atom:

> Does not it seem incongruous to be discussing the humanness of physics? For if one subject would appear to be lacking the quality of humanness, it is physics. Physics is characterized by precise measurement and abstruse mathematics, it is rigorous and austere; indeed it is

Fig. 1. "Caldron." Drawing by Angela Gonzalez

about as objective as a discipline can get. Yet, in spite of a prevalent belief that physics is cold and inhuman, a belief that it has to do only with thing, not people, a belief that its Faustian practitioners blindly follow the rites of scientific method to grind out a plethora of uninteresting facts . . . in spite of all of this, I am going to maintain that there is a quality of loveliness in the content and in the devices of physics, that it is beautiful creation which has meaning for man's view of himself and his place in the world, and that these qualities of physics can appropriately be discussed under the rubric of humanness.[4]

Richly illustrated with images from Greek mythology and drawings by Leonardo da Vinci and Angela Gonzalez, *The Humanness of Physics* acknowledged widespread critiques of institutionalized physics research in the United States and challenged directly the Faustian myth as it reemerged during the nuclear age, the myth of the scientist making a pact with the devil. Wilson made use of any and all available resources, blending classical and romantic imagery to give physics a new and deeper image. His rhetoric reveals the degree to which he understood the public's real or potential influence on the laboratory and the course of physics research. The document also reveals the degree to which the lab was, and is still, positioned within a larger, more complex dialogue about the future of physics. The boundaries he inscribed favorably distinguish physics from other fields of science. More important, they make clear physics relevance and relationship to the larger culture.

Any attempt to understand the rhetorical boundaries that circumscribe and define the laboratory must acknowledge how these boundaries create "publics" and bring them into a relationship. In the case of the national laboratory system, it is clear that, although "the public" may not design the experiments or technology, their "participation" justifies the continued existence of large-scale physics research programs in fields ranging from weapons production to astronomy. The history of technoscience in the twentieth century demands that we acknowledge that nonscientists do relate, react, and contribute to particle physics (as do other sciences and bodies of knowledge). To see this, we must account for the impact of rhetorics that define, expand, or restrict the full spectrum of nonscientists' social action and influence. Without question the laboratory's publics represent a wide range of constituencies critical to the life of the institution. Such publics are frequently defined as taxpayers, whose dollars fund the national laboratory system. Some are also neighbors, whose home life is affected both positively and negatively by their physical proximity to the laboratory. Some are employees of the laboratory or one of its contractors. Others relate to the laboratory as a cultural, educational, or environmental resource. This wide range of perspectives should be regarded as a prism through which the larger meaning of contemporary physics research can be refracted. Far too often, however, these perspectives are simply ignored. When the influence of nonscientists is acknowledged at all, the full range of perspectives is too often reduced to that of a single monolithic public, defined solely in terms of the contribution of tax dollars or thought of as subjects for (or objects of) science education. Such assumptions oversimplify the complex and interesting ways that nonscientists make sense of the laboratory. They mask the laboratory's significance and meaning in the lives of nonscientists and obscure the degree to which outsiders can and do actually influence the evolution of the laboratory.

As the laboratory's many constituencies grapple with the facility, its technology, and its social meaning, they negotiate a relation with the lab and its work. This gradual accretion of publics eventually defines the lab's boundaries and gives meaning to its spaces and its work. For the scientists and staff who work at the laboratory, *publics* are often defined according to budgetary needs and educational and environmental concerns. The resulting citizen-subjects of the laboratory are cast in a variety of roles. They are learners. They are neighbors. They are taxpayers. For each of these categories, there is a corresponding role for a laboratory employee. For the student of science, there is an educator. For the neighbor concerned about environmental or social impacts of the laboratory, there is an individual charged with explaining the realities of risk and the many benefits of the laboratory. For the taxpayer, there is a person with a job created by the laboratory. And there is a technology, an amalgam of metal, wire, and silicon that somehow speaks of the productive purpose of the laboratory and its people. All these characters are bound together by the rhetoric of the laboratory, those narratives and images that define its past, present, and future.

Unfortunately for scholars, many histories and analyses of modern laboratories fail to acknowledge the layers of rhetoric that structure the evolution of such institutions. The boundaries of the laboratory are defined according to the needs of real and imagined publics and the possible dialogues with them. Such dialogue is imagined in the public documents of the lab and discourse of those inside and outside its boundaries. The nature of this dialogue, however, reaches well beyond the boundaries of traditional argument to encompass the cultural experience of the laboratory. Examining the laboratory's image and narrative provides clues to the constitution of the lab's publics. It also produces insights into conflict negotiation and the resolution of controversy involving the lab. It took a series of complicated social transactions involving multiple publics, government agencies, scientists, and technologies to make Fermilab a physical reality.

Sadly much scholarship to date has not accounted for the work of nonscientists in the evolution of this and other scientific institutions. This limitation stems in part from an unsatisfactory understanding of the meanings nonscientists attribute to science. In order to better understand the social construction of scientific institutions, we must develop a more nuanced understanding of the meaning of science in society. While many researchers and philosophers have produced work that attempts to explain the *value* of science to society, far fewer have confronted its *meaning*. Work that does address questions of meaning rarely engages public controversy about the role of science in society. These shortcomings may result from our long-standing focus on the public's relationship to the products of science rather than the process of science. It is easy to see how fields such as medicine and chemistry produce products and procedures that both enhance and constrain everyday life, but it is far more difficult to see how physics might have a comparable effect for any but a few select individuals. If we shift our emphasis, however, and focus on the point at which research is conceived, funded, and sustained through the largess of the taxpaying public, we begin to see the complex dynamics at work among publicly funded physics labs and the public.

To address these and other limitations to our current understanding, this book explores the meanings *and* relationships that emerge from the rhetorical construction of laboratory spaces. This study is grounded in five years of ethnographic and archival research, interviews with employees of and visitors to Fermilab, and rhetorical analysis of the texts that mediate communication among the laboratories and the public.

The remainder of this chapter provides a narrative history of Fermilab and the field of high-energy particle physics. It details episodes of public controversy and explores the recurring narratives that mark the history and evolution of Fermilab. Chapter 2 surveys relevant research in the science studies and public understanding. It details the theoretical and pragmatic contributions that might result from further rhetorical studies of the laboratory. Chapter 3 examines the discourses that frame the laboratory and the discursive influence of its directors, Robert Wilson and Leon Lederman. Chapters 4 and 5 track important shifts in the laboratory's rhetoric and detail the images and narratives that define Fermilab for its employees (chapter 4) and for visitors (chapter 5). The final chapter considers recent changes at the laboratory and ponders the past, present, and future role of rhetoric in the articulation of the laboratory's boundaries.

Science and Society in the History of Government-Funded Physics Research

The National Acceleratory Laboratory (later renamed the Fermi National Accelerator Laboratory) was officially commissioned on 21 November 1967, by an act of the Atomic Energy Commission (AEC) and President Lyndon B. Johnson.[5] Devoted to a single purpose and single technology, the laboratory was unique. Today it continues to herald its commitment to basic high-energy physics research in its mission statement: "Fermi National Accelerator Laboratory advances the understanding of the fundamental nature of matter and energy providing leadership and resources for qualified researchers to conduct basic research at the frontiers of high-energy physics and related disciplines."[6] Explained more simply as "unlocking nature's deepest secrets, and learning how the universe is made and how it works," the mission of the laboratory is nonetheless difficult to define for the average taxpaying citizen.[7]

Before the construction of the Large Hadron Collider (LHC) at CERN (Conseil Européen pour la Recherche Nucléaire, now known as the European Organization for Nuclear Research), which was scheduled to circulate the first beams in September 2008, the Tevatron at Fermilab was the world's most powerful accelerator. The accelerator is built in the form of a ring so that particles can be repeatedly accelerated and stored before collisions. Protons and antiprotons are accelerated in opposite directions in the same ring and are brought into collision at specified locations. There elaborate detectors are used to analyze the interactions by studying the complex spray of resulting outgoing particles. Other kinds of high-energy physics experiments, called fixed-target experiments, are also located at Fermilab. In these cases an accelerated beam hits a fixed target, and the resulting interactions are studied. Yet another type of experiment produces a beam of neutrinos that is directed toward a detector located deep underground in northern Minnesota in order to study the phenomenon known as neutrino oscillation. All of these experiments involve probing the fundamental building blocks of matter and the forces present in our universe. Thus the work of studying basic particles and their behaviors is spread across several experiments asking a variety of basic research questions: What is the universe made of? How does the universe work? Why do particles have mass? What are the fundamental forces?[8]

With a 2007 operating budget of 263 million dollars, the work of the laboratory does not come cheap. When capital equipment and construction costs are included, the total annual budget tops 300 million.[9] The figure is staggering when one considers the abstract nature of the work and how difficult it is to explain it to the general public. It is quite reasonable for members of the public to wonder how and why this laboratory came to be. To comprehend fully the evolution of such "big sciences" at Fermilab, we must begin with an adequate understanding of the complicated history of the national laboratory system and the institutions that support it.

Although much has been written on the subject of wartime physics, few have sought to untangle the complicated series of developments that led from the institutions of wartime physics to a system of government-funded laboratories devoted to both basic and defense-related research. It seems reasonable to conclude, although few have pursued confirmation of this claim, that if physics was first organized on a large scale in response to a perceived public need (that is, war), public discourse might continue to influence the evolution of a system devoted primarily to basic research. In a postwar world we may acquire some insight into this possibility by examining early debates about the purpose and oversight of federally funded research—as well as public controversies regarding the establishment and operation of Fermilab. Important debates have shaped the evolution of Fermilab and demonstrate how the boundaries around organized physics have changed in response to the interests and concerns of nonscientists.[10]

Competing Visions—The AEC and the National Science Foundation

Most historians of science mark the beginning of the postwar science initiatives with the 1944 correspondence between President Roosevelt and Vannevar Bush, then director of the White House Office of Scientific Research and Development. In a letter to Bush concerning the development of postwar science, Roosevelt opined, "New frontiers of the mind are before us, and if they are pioneered with the same vision, boldness, and drive with which we have waged this war we can create a fuller and more fruitful employment and a fuller and more fruitful life."[11] In his now-famous response, Vannevar Bush expanded on the frontier mythology invoked by Roosevelt and wrote: "The pioneer spirit is still vigorous within this nation. Science offers a largely unexplored hinterland for the pioneer who has the tools for his task. The rewards of such exploration both for the nation and for the individual are great. Scientific progress is one essential key to security as a nation, to our better health, to more jobs, to higher standard of living and to our cultural progress."[12] In this single document, Bush claimed both the past and the future of a nation for science. Despite his personal desire to return to prewar scientific institutions and private systems of funding, it can be (and has been) argued that Bush's invocation of the frontier marked a profound shift in the way science was to be articulated with respect to social values and national goals.[13] Bush reasserted and developed Roosevelt's earlier claim by linking science not only to the production of knowledge but also to the production of jobs and technology for use in larger social contexts. In this incarnation science became not simply a tool for defense or industry, but the very "salvation" of the nation. As Mary Midgley puts it, this tendency to elevate and empower science has a long history in Western cultures: "Science is seen as having a special kind of value to which we all owe allegiance. People who want to list the glories of our civilization are almost sure to list science—meaning primarily physical science—among them, along with art. And the special value of science, like that of art, is not supposed to reach only the few who produce it, but also the public which receives it."[14] Bush's speech focused on the redemptive power of science in a world rocked by a war made particularly horrific by the products of organized research in physics and chemistry. Faced with the need to reorient public expectations and quell emerging fears, Bush argued that science and its institutions would be the pathway to prosperity in the postwar world. If Americans were to conquer remaining frontiers, they would need science to do so. For those in the sciences, the Bush report signaled—for good or for bad—"a new relationship between science and government."[15]

The frontiers to be conquered and nature of the relationship between government and science was the subject of heated debate. As Daniel J. Kevles describes in *The Physicists,* Bush desired a return to private and philanthropic funding of basic research, but he knew that the success of the wartime model argued strongly for more government intervention. In an attempt to strike a compromise favorable to his perspective, Bush proposed a system of "federal aid to basic research and training, along with a civilian-controlled effort in fundamental defense research."[16] Bush's plan was designed to satisfy the industrial interests that had exerted a great deal of control over the research process in the prewar years. Among these companies were many of the industrial giants that had fueled the American technological revolution from its beginnings at the turn of the century. Bush hoped to create a system that would not impinge on the patent process, prescribe geographical balance, or require any amount of socially oriented research. Put simply, Bush wanted to create the most "private" system of public funding possible—a system run for and by the scientific elite with the cooperation of large corporations and the military. As Kevles noted, "Bush conceded that his program had to be responsible but . . . not responsive—to the President and Congress."[17]

Bush's vision of government support was not shared by all involved. A critique offered by the *New Republic* argued, "Research needs to be coordinated carefully and the projects should be selected in terms of our national necessities, and not the accidental interests of various scientific groups."[18] This line of reasoning was common among Bush's opponents and made clear the focus of the debate: the locus of control in the new, federally funded system. Many who argued on behalf of government intervention and funding did so because they believed that such a change would ensure that science developed in keeping with the national interest and would always be responsible to the government and its constituents.

In many cases these individuals preferred the model of peacetime science proffered by Senator Harley Kilgore, Democrat from West Virginia. Kilgore advocated support for basic research and training in "all the major fields of science" and a "civilian controlled" defense program.[19] As Kevles describes, however, there were several fundamental distinctions between Kilgore's vision and that of Bush:

> Kilgore called for the evaluation and, in the form of a recommendation to the President, planning of all federal research activities for a good social purpose. He also urged federal support of the social sciences, insisted that at least part of the money in all fields be distributed on a geographical basis, and proposed that the ownership of patents arising from federally sponsored research normally go to the government. . . . To carry out his program, Kilgore proposed the establishment

of a new federal agency, a National Science Foundation [NSF], which would be a regular part of the executive branch and under the direct control of the President.[20]

Kevles casts this struggle in terms of a choice between "democratic" and "elitist" models of control, both powerful metaphors that continue to shape debates about public funding in science. Kilgore's model emphasized equal distribution and participation by arguing strongly for research that reflected national interest and promoted geographic balance and development. Bush, in contrast, advocated funding the "best science," a decision-making structure that would inevitably favor existing research facilities on either coast, the same facilities that were often developed by and continued to be associated with large industrial interests or prominent universities. Kilgore's proposal, particularly its call for centralized control and responsiveness, offered a model in which those in charge of scientific and technological development would be accountable to the American public for their work. Bush, in contrast, felt strongly that decisions should remain in the hands of those with the most experience. He argued that if trusted and left unhindered by unnecessary oversight or restriction on practices, government-funded science would evolve naturally toward the national interest in time. In the end Bush attempted to seize the rhetorical ground with the report *Science: The Endless Frontier*. As Kevles relates: "At the opening of the atomic age, the revolutionary needs of national security had joined the older requirements of economic development to force an end to what had long been, de facto, a federal policy of laissez-faire in physics. Bush aimed to enlist the nation's resources, through a politically elitist mechanism, to satisfy the scientific community's traditional goal of advancing the best science . . . a politically conservative interpretation of what the end of laissez-faire was to mean in postwar America for the Los Alamos generation and its science."[21]

What emerged in the coming years was a battle over the meaning of "public." "Science in the public interest" did not necessarily demand the same degree of visibility or transparency as "public science." The battle between these two perspectives continued for many more years and was joined by other influential individuals from the government and the sciences. The period from 1944 to 1947 was marked by persistent attempts to negotiate a compromise in the face of Bush's unwillingness to see science as subordinate or responsive to the federal administration. Much of the discourse focused on the proposed May-Johnson bill. This legislation would have structured nuclear research in terms favorable to Bush's vision. This bill met with much opposition from within the Truman administration because, following Bush's model, control within this system would rest with a part-time commission that would be only marginally responsive to the president.[22] Don K. Price of the Bureau of Budget and James R. Newman of the Office of War and Mobilization found this structure not only unpalatable, but potentially dangerous.[23] They felt that little attention was being paid to the developing relationship between the government and science and that, if left unchecked, control of important scientific matters might come to rest in inappropriate and irresponsible hands. Writing about Bush's ideas many years later in his book *The Scientific Estate,* Price asserted that "hardly anyone stopped to ask the fundamental question: how is science, with all its new power, to be related to our political purposes and values, and to our economic and constitutional system?"[24]

Price and Newman were not the only people to call attention to this problem. Many scientists, particularly those with wartime research experience, shared their concern. Kevles notes that, while notable physicists such as Ernest O. Lawrence, Enrico Fermi, and Robert Oppenheimer eventually endorsed the May-Johnson plan, the "rank and file" were deeply concerned about the degree of military control the bill proposed.[25] Despite repeated attempts to broaden the categories that defined federally funded science, the debates about "public interest" were inextricably bound to the history and proposed future of nuclear physics and its relationship to the weapons complex. The scientist's protest eventually developed focus, structure, and prominence with the creation of the Federation of American Scientists and the *Bulletin of the Atomic Scientist* in 1946.

In that same year, many scientists joined forces with Price and Newman to support a alternative bill developed by Price and forwarded by Brien McMahon, Democratic senator from Connecticut. Kevles characterized McMahon's proposal as "an atomic energy program civilian in control, liberal in purpose, and responsive to the political system."[26] Once modified to give the military a "significant voice in the atomic energy program," the bill was passed. It created a full-time, civilian Atomic Energy Commission "whose members were to be appointed by and responsible to the President."[27] The significance of presidential appointment is not to be overlooked. At this time the White House perspective on science was dominated by the highest-ranking members of the physics establishment. Even with modifications permitting military involvement, the passage of the McMahon bill signaled the power of the scientific elite to reshape a research infrastructure originally designed exclusively for defense interests under the banner of "public interest."

The creation of the AEC did not end the debate regarding the influence of the military on scientific research. The creation of the National Science Foundation, a funding agency independent of military control or interest, marked the next significant development in the relationship between institutionalized science and the larger culture. The NSF further expanded the political power of scientific elites. Bush envisioned a foundation that would serve the needs of scientists and "award grants, contracts, and fellowships to sponsor training and fundamental research" without the same demands for public accountability.[28] Bush's proposal, put forward by Senator Warren Magnuson, did not address issues of geographic balance, patent rights, or, most important, the degree of responsibility to the president. Once again the debate was cast in terms of public versus private control—Bush continued to advocate part-time administration on the part of private citizens (primarily scientists) whereas the Kilgore model of the NSF advocated a full time director appointed by the president.[29] While the Truman administration was not openly hostile to the idea of creating the foundation, the Magnuson bill was vetoed by Truman on the advice of Price, who questioned the wisdom of giving control of the National Science Foundation to the very people who would receive the grants.[30]

As Kevles details, a compromise was eventually struck that mandated shared leadership—between a director to be appointed by the president and a part-time civilian board made up of "spokesmen for the nation's leading institutions of academic and industrial science."[31] The inclusion of scientific and industry leaders on the board contributed to the institutionalization of "elites" in the governance of fundamental research in the United States. Kevles argues that this period of push and pull between Bush's and Kilgore's respective visions of the relationship between science and society would set the stage for what was to come in the next decades:

> In 1945 liberals like Kilgore may have wanted a federal program balanced between civilian and military needs, between big business and small business, between leading and less developed universities, between the welfare of science and the welfare of the nation. By the early 1950s the outcome was quite different: a program for the physical sciences dominated . . . its dispensations [were] concentrated geographically in the major universities, its primary energies devoted to the chief challenges of national defense and fundamental physics. . . . The result was produced not by design but by a combination of demands and defaults, including the demands and defaults of a new political power group, the nation's physical scientists. Whatever the cause of the outcome, it constituted a revolution in the relationship of American physicists to their society and their government.[32]

As the forthcoming chapters illustrate, the battle for control that Kevles describes is still alive and well in contemporary physics, albeit in transformed ways, and still provides the discursive frame for today's debates over the appropriate means of funding basic physics research in the United States. The debate centers on competing definitions of public interest—one broad, one narrow—that would shape the evolution of the science-society relationship. These competing interpretations are manifest in the scientific institutions and practices of a centralized system of laboratories that embody both the strengths and the instabilities of public science.

National Laboratories and the National Interest

By the spring of 1950, several of the wartime facilities had been declared permanent national laboratories by the AEC. While debates over control continued, the wartime contribution of physicists had provided a powerful justification for the peacetime expansion of research laboratories. In the years following the war, Argonne and Oak Ridge developed reactor research, while Los Alamos in New Mexico focused on weapons work. The Berkeley lab continued and expanded the accelerator-based physics research begun under Ernest O. Lawrence. These programs ranged the territory from nuclear medicine to civil engineering and emphasized both research and instrument design.[33]

Berkeley's long association with particle physics and the design and construction of particle accelerators epitomized Bush's vision of the endless frontier, particularly to the extent the program emphasized both scientific and industrial interests. In the early years of the Berkeley Radiation Laboratory (Rad Lab), research interests and technological discoveries ranged from hydroelectric power to medical therapies.[34] As Seidel claims, the blurring of the scientific and the technological was the hallmark of the lab's success. The laboratory accomplished this in part by resisting traditional methods of performing basic research. Instead of developing technology designed to test fundamental theories, they often pursued technological questions to their ultimate theoretical ends. Lawrence, the laboratory director, sought out energetic, young workers who shared his interest in the relationship between engineering and physics. This model of integrated, interdisciplinary science was central to Berkeley's success as a laboratory, and years later Robert Wilson would try to recreate the Berkeley environment within the single-purpose laboratory the eventually would become Fermilab.

On the opposite coast, I. I. Rabi and Norman Ramsey sought to create a radically different environment at Brookhaven Laboratory (BNL) on Long Island. Located on the site of a former army camp, Brookhaven worked to outrun both its physical and scientific ties to the military by focusing on pure forms of physics research. Work was conducted in both nuclear and accelerator research utilizing a graphite reactor and a synchrotron.[35] Brookhaven was the centerpiece of the northeastern research region and operated under contract from the AEC by Associated Universities, Inc. (AUI), a "consortium of nine universities."[36] BNL pioneered the consortium model of management. Unlike the Berkeley labs, where research was managed by a single university, the consortium permitted the coordination of power and the influence of staff and resources across several institutions. The consortium-contractor also provided an extra layer of insulation between the AEC and the laboratory in question. Many of those involved in the management of the labs felt this buffer would lessen the impact of government restrictions and regulations that they found troublesome. It worked also to distance the laboratories from the demands of public interest. Public concerns could be raised, but contractor management ensured that academic scientists would have significant influence on decisions about the nature and priorities of laboratory research. As the laboratory system grew, the consortium model would be replicated and refined. In the Midwest those advocating for an accelerator in the central region formed Midwestern Universities Research Association (MURA). Eventually Fermilab's management would rest with a similar group, University Research Associates (URA).

Bolstered by strong track records of successful research and effective management structures, Berkeley and Brookhaven led the way in the push for increasingly powerful accelerators. The production of mesons in the Berkeley synchrotron provided powerful justification for building newer and more powerful accelerators. In 1948 the AEC authorized the construction of a 2–3 billion electron volt (BeV) accelerator at Brookhaven and a 6–7 BeV machine at Berkeley.[37] Approval of these machines marked the starting point of large scale accelerator physics at the national labs.

By 1954 these machines were completed and operating. As the machines grew larger, more physicists and larger collaborations were needed to run them and perform experiments. It quickly became clear that technological expansion required a corresponding expansion in education to enable the production of new physicists. In this sense accelerators accelerated more than particles—they accelerated physics. Accelerator research was a boon not only for experimentalists already at work in the field, but also for university departments charged with supplying the next generation of scientists.[38] The success of machines was measured by the rapid-fire discovery of new particles, and in 1955 the AEC upped the experimental ante yet again by authorizing construction of a 25 BeV machine at Brookhaven.[39] The proliferation of new particles and increasing focus on accelerator technology specifically designed for this purpose resulted in the fundamental restructuring of the discipline. As Kevles describes: "The powerful accelerators built to probe nuclear forces by producing mesons were appropriated to the production of elementary particles, and gradually, the study of elementary particles emerged out of nuclear physics as a field unto itself."[40]

The development of such machines undoubtedly drove increases in basic physics research funding during the 1950s, but, as had proved true in the past, it was a wartime mentality that provided a particularly effective backdrop for this expansion. Greenberg argues that the logic of the cold war, particularly claims that the Soviets were in a race to "overtake American science," provided the incentive needed to build bigger and more powerful machines that would demonstrate superiority of the United States in the field of physics.[41] Such rhetoric drove the expansion of reactor research, as the development of nuclear power was forwarded as the alternative to the proliferation of nuclear weapons in Eisenhower's "Atoms for Peace" plan. Dramatically different technologies, reactors, and accelerators shared space in the social imaginary, each capable of symbolizing American supremacy in science. Thus, when Wolfgang Panofsky suggested a linear accelerator to be built in Palo Alto, California, and operated by the laboratory at Stanford University, the proposal needed to be related to cold war concerns in order to gain authorization.

In 1958 an advisory panel of physicists pointed out that the Soviet Union proposed to build a 50-billion-electron volt synchrotron, a machine twice as energetic as the most powerful budgeted for thus far in the United States at Brookhaven. The panel also endorsed the proposed Stanford linear accelerator. In May 1959 President Eisenhower announced at a symposium on basic research that he would ask Congress for the 100 million dollars necessary to build the Stanford machine. American progress in the field of high-energy physics, he declared, was vitally important to the nation.[42]

Despite persistent efforts to distance itself from the military-industrial complex and insulate its work from government oversight, the physics community found itself once again reliant on the discourse of war and American superiority as justification for its work. This complicated relationship between science and cold-war rhetorics would prove significant in the development of both SLAC and later Fermilab. Panofsky, an outspoken opponent of militarism and nuclear proliferation, countered governmental rhetorics by cultivating a strong sense of identification between Stanford University and the accelerator complex located in the hills several miles west of the campus.

Despite the early efforts of Panofsky and others, American science continued to grow in the shadow of the cold war, a competition now defined in scientific and technological terms. The United States responded to every Soviet advance with an initiative of its own, the development of the Apollo space program being perhaps the most notable example. Underlying this pattern of interaction was a corresponding expansion of the claims made by scientists. As science and technology became measures of success in the cold war, researchers drew on their newly found prestige and attempted to identify themselves once more with peaceful work. Kevles describes the expanding claims and influence of American physicists: "It was a time when Americans ranked nuclear physics third in occupational status . . . when physicists, among other scientists, were identified not only as the makers of bombs and rockets but as the progenitors of jet planes, computers, and direct dial telephoning, of transistor radios, stereophonic phonographs, and color television."[43]

Unlike in earlier decades, science often usurped engineering in the imagination of many publics. By claiming responsibility for such technological innovations, physicists generated public prestige and highlighted the importance of centralized science in cultural discourse. Scientific and technological achievement became synonymous with American achievement. In this way scientists began to regard their instruments and work as something more than research, as "monuments of Big Science—the huge rockets, the high-energy accelerators, the high-flux research reactors—symbols of our time."[44]

The degree of "establishment" and power enjoyed by scientists and scientific discourse was, however, disconcerting to some.[45] As Kevles and Price both have described it, the Eisenhower administration, despite having overseen a four-fold increase in research and development spending, was uneasy about the implications of unchecked growth in the scientific sector.[46] While Eisenhower's primary concern was the expansion of the military-industrial complex, he also warned of the dangers of surrendering decision making to science, of the "danger that public policy could itself become the captive of a scientific-technological elite."[47] This warning foreshadowed the paradoxical relationship between scientists, particularly those engaged in basic research at national laboratories, and the public during the 1960s. As Price argues, science had become "an establishment, in the old and proper sense of the word: a set of institutions supported by tax funds, but largely on faith, and without direct responsibility to political control," and this newfound status was suspect in the eyes of many.[48] The following years were marked by discourse about unnecessary or excessive expenditures and a "decidedly high standard of academic living."[49] Rumors were afoot about scientific conferences being held in glamorous locations and money being spent on salaries rather than research.

All of these factors combined to create an atmosphere where increased oversight and scrutiny of government-funded research was inevitable. This was the atmosphere into which the Ramsey report, the 1963 document advising the AEC about the construction of the "next" major accelerator facility, was received. Cold war conditions would shape much of the early decision making that led to the founding of Fermilab. For many the solution was thoughtful restructuring of the relationship between the government and scientific institutions. The Ramsey report, for example, acknowledged some of the government's concerns by prioritizing projects and sending a clear signal that physicists understood that government funds were not inexhaustible. "On time" and "within budget" became a mantra within the system, as Lyndon Johnson reminded physicists at the dedication of the Stanford accelerator.

Many scientists, including those involved in the drafting of the Ramsey report, were guided by what Price describes as "a faith in the political rationalism of the Enlightenment"[50] and thus allied themselves with neither political conservatives, who had little use for science, nor radicals, who felt science was central to the political system. As Price related in 1965: "Even the strongest critics of the government and its scientific policies—for example, many of the contributors to the *Bulletin of the Atomic Scientists*—are suprisingly traditional in their approach to the political system. They may question the capacity of our representative institutions to cope with the scientific revolution, but they tend to propose as remedies more international good will and cooperation, adequate scientific education of political leaders and the electorate, and unbiased scientific advice for members of Congress."[51]

These attitudes regarding the intersection of science and politics and growing tension between the two would not evolve a great deal in the coming years. Many of the above described remedies, designed to repair injuries to the relationship between science and society, were perpetuated in and by the discourse of the national laboratory system. When faced with challenges from outside, these institutions often retreated to their Enlightenment faith, expressing incredulity at the irrational concerns of the government or the taxpaying public, and vowed to educate the public better and bring legislators into contact with scientists who could explain and justify their requests. Educating the public in science was in fact inseparable from pleas for support of elite science.

The National Accelerator Laboratory and Its Changing Political Context

From the time of its founding in 1968 to its prominence following the discovery of the top quark in 1995 and on to the present day with questions about a future role in particle physics research, Fermilab's evolution as a research institution has been strongly influenced by competing cultural narratives that attempt to define what it means to do science in the public interest. The context sketched in the previous sections is important to bear in mind. As the previous discussions detail, the laboratory was created at the height of the cold war, when claims regarding the necessity and promise of big science and research were widespread and influential. However, such faith in the promise of science was soon tempered by sixties-style debates about the scope and social power of science, its practitioners, and its institutions. With enthusiasm for science counterbalanced by skepticism, the recommendations of elite panels did not always translate into the support required for a facility to be approved and funded. The National Accelerator Laboratory was one such recommendation.[52]

Founding Fermilab

Proposed as the home of a new 200 BeV machine strongly recommended in the 1963 report of the Ramsey committee, the NAL marked a significant shift in the rhetoric of the national laboratory system. Building on the example set by its predecessor, SLAC, the NAL drew even more heavily on the trope of "basic research." This rhetorical strategy still evoked visions of the "new frontier," and yet it also raised questions by those committed to competing interpretations of public interest, questions to which the project was forced to respond. In the face of queries from critics and budget-conscious legislators, some physicists adopted a decidedly separatist posture intended to distinguish the new laboratory from the weapons labs and other facilities associated with the military-industrial complex. Their strategy of drawing from a boundary between weapons labs and research labs reinforced calls for scientific control of new facilities. The research labs were thus reinterpreted as cultural rather than governmental institutions. Shifting the rhetorical frame not only redefined the mission of the laboratory; it reorganized the very standards by which the necessity and value of the laboratory should be evaluated. When asked by a congressman what the NAL would contribute to the national defense, Robert Wilson responded defiantly, "This new knowledge has all to do with honor and country, but it has nothing to do directly with defending our country, except to help make it worth defending."[53] Wilson's characterization of the laboratory's sociocultural value foreshadowed the emphasis on the technological sublime that followed, and the issues form one of the key themes in our inquiry. Even more important, it indicated the degree to which the physicists were cognizant of the need to reconstitute the public image of their field in response to persistent critiques that questioned the wisdom of public funding for basic research.

This process of reinvention was complicated by regional and institutional tensions within physics research. The creation of the NAL came at the expense of the machine already being developed by Midwest University Research Associates. The Ramsey report did not suggest the termination of the MURA accelerator but recommended that it be continued only if it did not interfere with the field's development of the 200 BeV design. This was interpreted as the "kiss of death" by many of those involved and has long been regarded as the pivotal act sealing the fate of the machine that midwestern physicists had lobbied long and hard to obtain.[54] In the end the termination of MURA's project would prove to be a critical factor in site selection for the NAL. Midwestern physicists were increasingly concerned about a lack of facilities in the middle region of the country and what they considered to be the provincial management of Brookhaven National Laboratory on the East Coast and Lawrence Berkeley National Laboratory (LBL) on the West Coast.[55] From the perspective of midwestern researchers, proposals submitted by researchers from other areas than these were often given low priority at LBL and BNL. The creation of the NAL reflected not only this tension in the field but also the rebalancing of power within the physics community. Leon Lederman supplied the key phrase to symbolize the shift, arguing that the new facility should be a "truly national laboratory" where access to the accelerator would be decided based on the merits of the research proposal rather than the institutional affiliation of the scientists involved.[56] Given this new, more democratic model of organization and the de facto cancellation of the MURA facility, midwestern researchers felt justified in calling for an accelerator that would be centrally located. It was further suggested that the new facility be managed by a consortium of research institutions that included representation from midwestern universities. The creation of the University Research Associates (URA) in 1965 satisfied the latter proposal. But only an untested site search would answer the former.

While the site search committee considered the advantages of a laboratory that was geographically central, there were other nonscientific factors that would contribute to the decision. The mid-1960s were a period of economic and social upheaval and these larger cultural concerns had a significant impact on the decision-making processes surrounding the laboratory. Advocates for locating the facility in California were associated with Lawrence Berkeley National Laboratory and perceived to have a natural advantage in that their design made use of existing laboratory infrastructure and facilities. This efficiency came as no surprise as the California contingent, headed by Edward Lofgren, had had a great deal of lead time to plan and propose a site-specific design. The design they were developing, however, was surprisingly expensive and, in the opinion of Robert Wilson and others, unnecessarily conservative. Seeing an opportunity to influence the design and siting decisions, Wilson offered an alternative design that was simpler and less expensive. This less-expensive design provided an alternative sufficient to influence deliberation among legislators increasingly worried about the growing expenses of the Vietnam war. Although the Berkeley design remained the "official" choice for quite some time, Wilson's technical rhetoric opened the door for another design and another location.

A further complicating factor could be found in the racial tension that wracked the nation at the time. As the site selection process progressed, it became apparent that housing laws might affect the decision-making process.

Advocates of open housing saw the site selection process as an opportunity to raise public awareness of discriminatory practices and strengthen the case for federal open-housing laws. Given Illinois's history of resistance to open-housing legislation, the state's finalist sites, Weston and Barrington, became targets of strong protest on the part of the NAACP and African American activists from the Chicago area. Illinois had failed to pass open-housing legislation on two occasions, whereas Denver, a competing site, already had such laws. Many felt that a "frontier" facility should reflect social as well as scientific frontiers and should under no circumstances be located in a state that that was so clearly resistant to social progress. According to Norman Ramsey, some members of the committee believed that voting for the Illinois site might hasten the passage of such legislation in the state of Illinois and ultimately benefit the large African American population in the greater Chicago area. The debate grew increasingly polarized as the site selection committee moved toward a decision. The NAACP and Martin Luther King, Jr., threatened to stage a protest at the laboratory and urged reconsideration of the decision criteria. The Barrington site was eventually chosen, but to the great surprise and embarrassment of the site selection committee and the AEC, the affluent Barrington community declared that they did not need or want the facility. On 16 December 1966 the selection committee eventually named Weston as the new site for the NAL.[57]

A curious narrative has evolved in the physics community that actually credits this site selection process with turning the tide of open housing legislation. For many, the influence wielded by the Republican Senate minority leader, Everett Dirksen, in the open-housing debate, particularly his reversal on the issue between 1964 and 1968, suggests that he may have been swayed by the opportunity to secure the research facility for his Illinois constituents. It is widely believed in the physics community that Lyndon Johnson made good use of both the potential economic impact of the proposed facility as well as the surrounding controversy in order to persuade Dirksen to cast his vote in favor of open housing. Ned Goldwasser, the laboratory's first deputy director remembers it this way: "My feeling is that President Lyndon Johnson made the decision at least in some measure as a tradeoff with Senator Everett Dirksen of Illinois. . . . What surprised everyone was that Everett Dirksen, who had a long record of strong positions against anything in the nature of open housing, withheld his vote to the very end, and then he cast his vote in favor of the bill to break the tie. My own feeling is that this had something significant to do with the choice of Illinois for the site."[58]

Today theories abound regarding the decision to locate the facility on a greenfield site in Illinois. Most scholars and physics insiders agree that in the end economic, social, and political factors, not physics criteria, decided the issue.⁵⁹ The timing of the open-housing decision suggests, however, that the influence of the site selection battle may be overestimated. Dirksen voiced his objections to open housing in 1964, leading a successful Republican filibuster of the proposed legislation. Although the site recommendation in 1964 may have played a minor role in Johnson's attempts to deal with Dirksen's objections, the Senate vote in favor of the Fair Housing Act did not take place until March 1968. Dirksen's change of heart may be more accurately attributed to significant changes in the proposed legislation itself, namely the removal of any significant enforcement mechanisms from the language of the bill. It seems more likely that this compromise, rather than the promise of a new physics laboratory, secured Dirksen's vote and eliminated the threat of yet another Republican filibuster.⁶⁰

The fact that the story of the Fair Housing Act has been persistently and creatively rewritten by Fermilab insiders is reflective of the identity and values of the institution. The belief that site selection was influential in turning the tide in favor of the act is strongly linked to the lab's early attempts to reach out to minority communities through mentoring, education, and job-training programs. As discussed later, even the laboratory's earliest rhetorics situate science in relationship to social justice. The year of the lab's founding was marked by the Tet Offensive in Vietnam and the assassinations of Martin Luther King, Jr., and Robert Kennedy. Wilson's characterization of the lab as an institution that "would make the country worth defending" came less than a year later. As Goldwasser would later recall: "Those were some of my interesting days. I met with leaders of the Urban League and the National Association for the Advancement of Colored People, but I also met with leaders of the Black Panthers, as well as with gangs in Chicago. I told them what our intentions were, and asked them to give us ideas about how we might proceed."⁶¹

Wilson, Goldwasser, and the community of scientists gathered at the lab envisioned an environment that would be an example not only of scientific values but also of the emerging human rights consciousness of the time. This vision was, in all likelihood, both a reflection of their idealism and also a strategic response to criticisms that had been leveled at the laboratory and the physics community during the lab's earliest days. As a demonstration of their commitment, Wilson and Goldwasser drafted a human rights policy that is a pithy interweaving of the scientific and social values of the sixties and seventies as well as a direct response to the protest raised during the site selection process: "In any conflict between technical expediency and human rights, we shall stand firmly on the side of human rights. This stand is taken because of, rather than in spite of, a dedication to science. . . . Our support of the rights of members of minority groups in our Laboratory and its environs is inextricably intertwined with our goal of creating a new center of technical and scientific excellence. The latter cannot be achieved unless we are successful in the former."[62] Wilson and Goldwasser intended this statement to be more than mere policy. It was to be a public demonstration of what the laboratory could mean to the community and the country if properly conceived and understood.

Operating the New Laboratory

While the choice of an Illinois site was regarded as a victory for midwestern physicists, coastal researchers, particularly those from Berkeley, were angry. After the Weston site was chosen, Edward Lofgren refused to head the project despite the AEC's consistent support for the Berkeley design.[63] Tightening budgets and Lofgren's refusal to build the "Berkeley accelerator" at the Weston site opened the door for Robert Wilson's more radical approach to accelerator design and laboratory management. Wilson was offered the director's post in January 1967 and accepted on 7 March 1967, after receiving assurances from the more conservative members of the AEC that he would be allowed to execute the design and construction according to his own sensibilities. On 21 November 1967, Lyndon Johnson signed the bill officially authorizing the National Accelerator Laboratory.[64]

Having chosen the site and the director and having agreed in principle to a design, the AEC left day-to-day management of the project to Wilson. His primary concern was to gather together a community of young physicists who shared his passion for the design and building of accelerators and his vision of a scientific utopia. To accomplish these goals, he organized workshops on accelerator design as a method of recruiting interested and talented physicists. These workshops functioned as a gathering place where physicists explored the technical and experimental possibilities of the new machine and ironed out design details. These early workshops, housed in makeshift quarters in Oak Brook, Illinois, became an incubator for the burgeoning community.

Fermilab flourished despite early challenges. Congress authorized the first funding for the facility in 1968, and groundbreaking for the first stages of the cascade of accelerators began late that same year.[65] Wilson's passion for simplicity and efficiency was expressed in his management style as well as his designs. As Hoddeson and Westfall argue, "He built the accelerator using an approach previously employed for smaller machines that he tailored to accommodate his own aesthetics, including a preference for a holistically-conceived, spare design."[66] Hoddeson and Westfall attribute this approach to Wilson's experiences with Lawrence at the Berkeley Radiation Laboratory, where "accelerators were built at breakneck speed, using whatever resources were available."[67] According to Rich Orr, this system worked well through the early phases of construction, but required substantial revision when problems arose with the delivery of beam to the main ring in 1971.[68] In response to these problems and the fear that the ring would not be completed on time and under budget, Wilson appointed three managers and a corresponding set of commissioners to oversee the problems of the accelerator. This new system was effective, and by 1 March 1972 the first 200 GeV beam passed through the main ring.[69] In 1974 the facility was renamed Fermi National Accelerator Laboratory in honor of University of Chicago physicist Enrico Fermi.[70]

Funding shortages continued throughout the seventies, even as the 200 GeV accelerator was functioning and producing interesting physics results that included the discovery of the bottom quark in 1977. The planned upgrade to 600–1,000 GeV was put in question because of both technical and budget concerns. Discouraged by the federal government's initial rejection of funding for the upgrade, Wilson retired in 1978 and was replaced by Leon Lederman, a physics insider accustomed to the push and pull of the government funding process. The shift to a Republican administration three years later was initially favorable for the national laboratory system, and Lederman's term as director was marked by renewed support for technological upgrades.

In 1983 the first beam was sent to the "energy doubler," a second ring of super-conducting magnets designed eventually to double the power of the accelerator. The doubler was soon renamed the "Tevatron" in anticipation of the fact that the machine would measure beam energies in terra electron volts (TeV) instead of giga electron volts (GeV). The beam energy reached 800 GeV in 1984 and 900 GeV in 1986. The increase in beam energy was accompanied by the initiation of colliding beam experiments, a substantial and significant change in the practice of experimental particle physics. It is somewhat ironic that the focus on technological improvements during this era meant that less actual physics research was being done at Fermilab and in the United States generally. Thus, during this period, many of the significant discoveries were made at facilities overseas. In 1983 CERN, the premier facility in Europe, announced the discovery of the W and Z particles. While Fermilab continued to earn recognition for its substantial technological achievements, its push to the edge of the "energy frontier," and its work in magnet research and testing and detector development, it seemed to be losing the battle in terms of particle discoveries.[71]

Despite the drought in discoveries, the lab fared better during the Reagan era than some branches of basic research. Wartime associations between physics and the national defense seemed to reemerge with Reagan's widening of the cold war, and physicists seized this rhetorical opportunity to call for the construction of a very big superconducting supercollider, a machine that would have little to do with defense but everything to do with keeping the United States at the forefront of particle physics research.[72] As the eighties came to a close, those at Fermilab were looking forward to the superconducting supercollider and to a long future at the energy frontier. In anticipation of a shortage of physicists needed to staff such a large facility, Lederman began a number of education programs designed to interest young people in scientific and technical careers and to prepare the next generation of experimentalists who would be needed at the new accelerator.

Fermilab in the Era of the SSC

The end of the cold war and changes in the White House, however, fostered increased skepticism about unquestioned support and "automatic funding" for big science projects. Public discourse about science took on a new, confrontational tone that would have serious implications for the future of Fermilab.[73] Requests for expansion or upgrades of existing facilities met with increased scrutiny as control over the future of high-energy physics research gradually shifted from the scientific community to a federal government determined to demonstrate its attentiveness to the public's concerns about immense budgets and deficit spending. For example, when the superconducting super collider (SSC) was initially proposed in the early 1980s, many argued that Fermilab was the logical site for the facility as much of the necessary infrastructure and R&D facilities were already in place. Public controversy over locating the facility in Illinois in combination with governmental pressure to "spread the wealth" of such projects led Congress to advocate for an open site contest for the SSC. While a Fermilab-Illinois site progressed to the final round of the selection process, DOE research into community attitudes at the proposed locations revealed an unexpected pocket of resistance in the Batavia area.[74] Community reluctance in Batavia, combined with strong political pressure in favor of siting the supercollider in Waxahachie, Texas, effectively eliminated Fermilab as a contender.

In 1989 Leon Lederman retired. John Peoples was appointed director of the laboratory. The development of the SSC was continuing apace, and, despite Fermilab's failure to prevail in the site selection process, the laboratory was deeply invested in the success of the project. Indeed Fermilab physicists were involved in many aspects of the SSC design, most notably the development of superconducting magnets for the accelerator. In addition to its contributions in preparation for the next machine, the lab was also focusing on the first experimental results from the Tevatron. Significant upgrades were made to the existing facilities during this period including the commissioning of a second large-scale collision detector named "D-Zero," a new 400 MeV linear accelerator, and groundbreaking for the main injector, an injection ring designed to double the potential research capacity of the laboratory by allowing for simultaneous operation of collider and fixed-target experiments. The expansion of Fermilab was intended to keep the United States in firm command of the energy frontier until the SSC was completed.

Europe's success in the early 1980s provided the momentum for CERN to propose its own version of the SSC, the large hadron collider (LHC). This European initiative led to the inability to secure international funding. When the collapse of the Soviet Union led to the diminishing persuasiveness of cold-war-era rhetoric, that, along with the lack of international funding, indicated that the fate of the next machine in the United States seemed to be sealed. In 1993 the SSC was defunded by Congress.[75] The cancellation of the SSC meant that Fermilab would remain, for the near future at least, the site of the world's most powerful particle accelerator and thus of the "energy frontier" for particle physicists.[76] Indeed the threat of reprisals in response to lost SSC jobs probably allowed Fermilab's many upgrades to survive several rounds of DOE budget cutting that had been necessitated by the SSC's voracious appetite for funds. Human talent also had to be reallocated among the labs, and, while it reaped no direct financial gain from the cancellation of the SSC, Fermilab fared better than many labs from the dispersal of talent that had gathered in Texas. Some physicists relocated or returned to resume their work in Batavia while others simply left the field.[77] Young scientists were particularly hard hit by the cancellation of the SSC, having been left without a facility at which to train and with no realistic prospects for jobs within the United States. As in the early 1980s, the country that had prided itself on braving the frontier of high-energy physics suddenly found itself looking to Europe for the "next machine." The prospect of living and working in Europe, but especially the prospect of being dependent on a European machine, was unpalatable to many American physicists and almost impossible to comprehend for some who had seen the field develop in the United States from the earliest days of Berkeley's Rad Lab.

In the midst of the shock over the unexpected demise of the SSC, therefore, the Fermilab community found reason to celebrate. After many years of searching, in 1994 experiments conducted with the Tevatron revealed the first direct evidence of the top quark, a long-sought building block of matter predicted by the standard model.[78] By 3 March 1995 researchers from both detector collaborations had confirmed their results and officially announced their discovery.[79] This groundbreaking achievement, in combination with a record-setting number of proton-antiproton particle collisions, allowed the laboratory publicly to declare its dominance in the field of high-energy research. Preparing for the main-injector upgrade necessitated shutting down the high-energy, fixed-target work. In 1996 and 1997, shortly after the collider run that produced the top quark, a final 800 GeV fixed-target run took place.[80] Following this run, the main ring was decommissioned with pomp and ceremony typical of the laboratory.

As the time of this writing, the main injector has begun operation following a lengthy commissioning period.[81] The laboratory has seen two new directors since John Peoples left shortly before the construction of the main injector began. The succeeding directors, Mike Witherell (1999–2005) and Pier Oddone (2005 to present), have worked to maintain community and public support for the laboratory during this important transition. In the past the testing and commissioning of the overhauled accelerator has not gone smoothly.[82] In the spring and summer of 2003, tensions ran particularly high as the beam could not be delivered at the energies promised to researchers. Media reports about trouble with the new technology, along with growing frustration among researchers lined up to use the machine, led the laboratory once again to promise experimenters and the public that it would deliver the technology "on time and on budget."

Fermilab continues to be shaped by recurring rhetorics that mark the progress of its research, the development of each new technology, and the nature of its relationship with the government and the general public. The discovery and measurement of the top quark and several new projects integrating astrophysics and particle physics did much to return the laboratory to the level of recognition it enjoyed in its earliest years. But, as in those years, the expectations were high, and meanings of the laboratory were contested. Thus, even in the warm afterglow of such a momentous and long-awaited discovery as the top quark, Fermilab must work hard to continue to justify its own existence. Recurrent congressional proposals to privatize the national laboratory system and dismantle the DOE require Fermilab to make its own case for continued governmental support with each new budget cycle. These actions only compound the effects of a string of management changes at the DOE and the implementation of objective performance standards as required by the Government Performance and Results Act of 1993.

As the main injector pulses along, it is difficult to predict its future as a "frontier" machine. The history of Fermilab detailed here is part of a complex tapestry that reflects those moments when the institution met its publics and was forced to negotiate or legitimate its worth to nontechnical audiences. That the public and its representatives have exerted and continue to exert influence in the development of high-energy physics is clear. The emergence of government initiatives to direct and monitor the course of energy research can be traced in part to conflicting accounts of public opinion—general support for scientific research coupled with increasing public concern about government accountability.

Predicting the future of physics funding is further complicated by the fact that the high-energy-physics budget is shaped not only by public opinion about science but also by the government's comparative assessment of issues pertaining to housing and veteran's affairs, the DOE's two closest and most salient competitors in the budgetary scramble for discretionary funds. In the end decisions that benefit or harm individual programs are often made on the basis of what the public feels holds the greatest benefit or importance for society. As Neal Lane, director of the National Science Foundation, stated in an address at the 1997 Fermilab Users Meeting, "science policy has become public policy."[83]

The following chapter offers a specifically rhetorical perspective on the history of Fermilab and the relationship between science and society. In this chapter I have set the field for an inquiry into the rhetorics of Fermilab and offered a narrative of its arrival and short career that both reflects these rhetorics and contextualizes them. By considering how the boundaries of the laboratory are constituted rhetorically, especially around "frontier rhetoric," as well as scientifically, governmentally, and technologically, we can better comprehend both its past and its likely future. Drawing on literature from science studies and rhetoric, chapter 3 presents theoretical and methodological resources to help us grapple with the importance of communication in the life of a laboratory. Communication and rhetoric are the primary means by which the laboratory constructs its identity for its publics and, when necessary, manages controversy. Furthermore the rhetorics that organize public perceptions of the laboratory also work to create the institutional culture that shapes day-to-day life for laboratory employees. Taken together, these internal and external rhetorics constitute the frameworks within which the laboratory operates and by which it argues for continued support.

Chapter 4 and 5 explore the effectiveness of these rhetorics. They are based on interviews of people working in various roles within Fermilab and with external audiences who visit the laboratory—or rather did in a more open environment prior to the event of 11 September 2001. That event reduced public access to the FNAL, seriously affecting the laboratory's boundaries. I end with a chapter discussing the current state and prospects of Fermilab and its rhetorical orientation.

⟯ TWO ⟮

Mapping the Boundaries of the Laboratory

I've loved science since I was a child and have long been fascinated with physics in particular. Growing up in Illinois, I knew about Fermilab. I knew it was a government lab. I knew it had something to do with physics. But I didn't know much more. I remember once driving by the gates and experiencing the mystery of the lab. I had so many questions. What went on in there? Was it classified? Was it dangerous? I would later learn that the answer to the first question was "high-energy physics." The response to the second and third queries was a resounding "no." At the time there was no physical boundary around the laboratory—no guard gate, no chain-link fence, nothing foreboding to prevent my entrance. And yet an unseen rhetorical boundary defined the space and drove my initial understanding of "inside" and "outside," "science" and "society," "me" and all those people inside the laboratory.

Today security measures and the pressures of urbanization have changed the laboratory's physical boundaries, raising new questions for anyone driving past the laboratory. Still, my first impressions of the laboratory stuck with me, particularly as I encountered others who had similar questions. Many people wondered just what went on inside the laboratory. Some grew worried as they speculated about the type of research being done. Why the mystique? Why the curious mix of attraction, awe, and fear? As a rhetorician, I pondered the persuasive dimensions of my own and others responses to the laboratory. Was I tapping into a localized reaction or a more generalized cultural understanding of science and technology? Did people respond this way because it was a government lab? Most important, was there any common ground that could be shared by those inside and those outside the lab?

Negotiating the Relationship between Science and Society

There is a wide-ranging literature that confronts, either directly or indirectly, the interaction between science and the larger culture. While differing on many points, scholars who study science commonly regard communication as a fundamental component of the production, translation, and public reception of scientific knowledge. The inescapable influence of culture on the practice of science makes itself felt in these studies. The myth is that science affects society, not the other way around. Work in the history of science studies and rhetoric of science, however, includes many compelling accounts of how society too shapes scientific practice and scientific institutions. Much of this work describes how science and scientists at various points in history have been influenced by public discourse in the development and presentation of their work.[1] By framing the public as an *audience* for scientific discourse, we can study the rhetoric associated with scientific discoveries and the development of scientific institutions. In turn this sort of study provides a reasoned analysis of the influence of *nontechnical* arguments and discourse on deliberations about science, scientific institutions, and science policy.[2] From this perspective the development of FNAL has been and continues to be influenced by public concerns and interests as well as ongoing active negotiation of the boundaries of physics research

The next question is how the public in fact views science and what scholars have said about that. In the following discussion I consider the benefits and limitations of previous research into the public's attitudes about science. I review boundary-work studies that address how the meanings of science and technology are negotiated in the larger culture. Lastly, because the performance of "scientific character" is particularly important to the history of Fermilab, my discussion includes studies that explore the rhetoric of scientists and how it is perceived by public audiences.

Public Attitudes, Interest, and Understanding

What little we do know of the public's understanding of and involvement in science has been generated primarily through survey research and other forms of large scale sampling.[3] Such research has provided scholars with a partial picture of what the public knows about science, the "facts" they have learned through exposure to various sources of scientific information, and their general attitudes toward various scientific projects and policies. A review of the past three decades of survey research designed to assess public attitudes toward science reveals interesting yet somewhat conflicting results. Survey data, collected biennially by the National Science Board (NSB) since 1957, indicate consistently high levels of support for science and technology. A substantial majority of the surveyed public feels that science does more good than harm, agrees that their lives have been improved through the application of science and technology, has high confidence in scientists and scientific institutions, and has high expectations for developments in science and technology in the next twenty-five years. In general Americans and Europeans express faith in science and support both basic and applied research. However, this generalized support breaks down when respondents are asked to rank science against other pressing concerns. Scientific research consistently ranks sixth or seventh behind other issues uppermost in the public mind such as crime, pollution, education, health care, the economy, and care for the elderly. Furthermore, although support remains strong, more recent surveys have noted an increase in the percentage of people reporting negative associations with science or a lack of confidence in the prospects for science and technology.

In addition to monitoring public support for science, these surveys have also sought to determine the degree to which the American public is attentive to and informed about science, effectively creating the category designation of "interested" or "attentive" publics. While a large percentage of Americans describe themselves as interested in science and technology, survey research has often distinguished between interest and attentiveness.[4] Since 1979, when attentiveness was first measured, the segment of the population considered attentive to issues of science and technology has remained stable at approximately 10 percent.[5] Inconsistencies arise, however, when considering this attentive public on a finer scale. Men, for example, exhibit generally higher levels of attentiveness than women but also attend to different issues than their female counterparts: men appear more likely to monitor developments in the theoretical sciences and technology, whereas women seem to attend to issues involving the application of scientific knowledge including medicine, health care, and the environment. These distinctions may indicate the influence of meaning derived through experience not accessible through survey methodologies. They must also be considered artifacts of the survey design.

The relationship between knowledge and attentiveness to or interest in science also poses important new questions. Fewer than a third of American adults consider themselves knowledgeable about science, and fewer still can pass a simple test of basic scientific knowledge. But this confessed low level of knowledge seems to have no effect on support or opposition to science. Furthermore certain issues, such as nuclear power and pollution, seem to have particular salience for the general public and command widespread attention despite persistent low levels of knowledge. These findings should call into question measures that automatically link knowledge levels to attentiveness and participation in public discourse about science. Close examination of the inconsistencies in the data would seem to suggest that individual opinions about science and technology are constituted through meanings that are based in interest and experience as much as from formal science education or even informal exposure to scientific issues through television, newspapers, museums, and magazines.

Survey research such as the National Science Board's science and engineering indicators provides a broad perspective on the information transfer between science and the lay public. These studies inform our understanding of the various sources of scientific information and expose gaps between knowledge and opinion formation. However, the procedures employed in these studies are not capable of accessing the qualitative experience and interpretation of science that is the actual formative context for such knowledge and opinions. Much survey research tends instead to frame public understanding in overly simplistic terms of proficiency and deficiency. But this common approach confuses the acquisition of prefixed knowledge with the understanding, judgment, and application of knowledge. Failure to distinguish between the two poles often serves to marginalize, rather than explain, the nonscientist's experience of science. It also obscures the roots of public debates about science, suggesting that adequate scientific education would eliminate public controversies about science and technology. There is a great deal of evidence suggesting this is not the case.

The inherent ambiguity of NSB survey results, combined with a tendency to reinforce existing interpretations of science, should provoke critical questions about the picture of public understanding of science drawn from the NSB and other survey studies. Similar concerns were expressed more than twenty years ago by Pion and Lipsey, who characterized survey work designed to measure public attitudes toward science as rather general and asserted that "public attitudes show many indications of being unusually vague, distorted, and superficial."[6] Among other things, Pion and Lipsey advocated research designed to explore "how [scientific] concepts are connected with the respondent's experiences" so that we might develop a more nuanced understanding of the meaning nonscientists attribute to science.[7] Although the number of studies attempting to address these concerns has increased, much work remains to be done.

If present research cannot fully account for the nature and quality of interaction between science and the public, we are left with intolerable gaps in our narrative of the history of the relationship between science and society and its impact on institutions such as Fermilab. It is tempting to oversimplify and envision idealized encounters between the publics called on to adjudicate the future of federally funded research. Some argue the process is straightforward: the public agrees to fund, through federal government programs and agencies, those research programs that serve the common good. Taken at face value, this would appear to be an instance in which the scientific community, the state, and the public work together to constitute a "scientific public sphere" and so decide the future course of science. However, as critiques of the public sphere detail, actual sociopolitical relationships are rarely this simple. Often the very discoveries and technologies that result from publicly funded research complicate and confound the process. The history of physics reveals substantial fluctuations in the degree and nature of involvement on the part of various public and government entities. At times the government has more or less successfully absorbed physics within the military-industrial complex. At other points physicists resisted this process and attempted to reclaim authority over the trajectory of their research. Peppered throughout this give-and-take history between physicists and the federal government are instances in which publics affected the course of development by either supporting or refusing to support particular institutions, technologies, or research programs. It would seem the historical narratives of particular cases trump any generalization we care to make.

Still, researchers often have attempted to explain these fluctuations. While some analyses seem to reinforce perception of an unproblematic and wise scientific public sphere, others depict a world in which the technical expertise dominates the decision-making process and ordinary citizens are simply taken in by the epistemic authority of science.[8] Neither explanation is entirely adequate. Close examination reveals a wide variety of public discourses about science that extends beyond issues of knowledge production. These discourses compete for attention in the popular imagination and thus generate dissension and controversy regarding what scientists—particularly physicists—have done for the common good.

Historical accounts of the development of basic physics research actually reflect how the relationship between science and its publics has been negotiated through rhetoric. Long-standing disputes about the goals of publicly funded research have often resulted from competing interpretations of the nature, practice, and social worth of various types of scientific research. These disputes create an environment that simultaneously promotes and obfuscates the participation of nontechnical, "outsider" publics. For example, stakeholders may be acknowledged by federal agencies even as their contributions are discounted on the basis of insufficient expertise. The conditions that enable nontechnical public participation are of critical importance to those interested in the rhetoric of science. We must better understand how, why, and on which occasions scientists and nonscientists alike violate the boundaries that consistently separate the technical and nontechnical public. If we focus attention on the boundary rhetorics of science and technology, we can better discern the nature of public participation and identify the obstacles that complicate and prevent such participation. The study of boundary rhetorics also furthers our understanding of the constitution of scientific organizations, learned professions, and various publics. Careful examination of their respective contributions helps us to map past and present cultural discourses of science and technology. Such knowledge may help future researchers to evaluate better the evolution of public controversies about science and technology and the effectiveness of various attempts to resolve them.

Boundary Work and the Development of the Relationship between Science and Society

Much of the work in the philosophy, history, and sociology of science has sought to articulate the boundaries of scientific method and practice. Early work in this area often sought to answer a deceptively simple question: "What is science?" These studies, philosophy of science in particular, explored the vast territory of scientific inquiry, differentiating the practices of science from other modes of scholarship and inquiry.[9] In doing so, they participated in the process of segregating science from society. Though this work provided much-needed perspective on how science looks and feels from the inside, it was and is somewhat less useful for understanding the emergence of *relationships* among scientific practices and institutions and many other dimensions of social life. To fill this gap, other scholars have focused on the very process by which science became "science" in language and in the social imaginary, how science in other words achieved its "aura and status." Examining the way in which science becomes meaningful in the larger culture, studies of this sort provided an avenue for understanding both the contributions of science and the controversies that frame and define the relationship between science and society.[10] This body of work acknowledged the influence of outsiders on the form and function of scientific discourses, but did not offer detailed accounts of precisely how nonscientists and nonscientific discourses contribute to the boundary work process.

A rhetorical perspective on boundary work focuses on the actual interplay among the discourses of scientists and nonscientists, challenging the idea that internal and external discourses are separate and that the public and private experience of science are somehow mutually exclusive.[11] Charles Alan Taylor's metaphor of the "scientific ecosystem" provides an excellent example of how rhetoric aids in understanding the complex and often paradoxical relationships among taxpayers, federal government, and the federally funded research institution.[12] Instead of regarding science as a series of independent discoveries, disciplines, or institutions, Taylor employs the ecological metaphor to encourage us to think of science as an integrated system developing within a larger social context that is sustained through discourse. The metaphor of the ecosystem can help to redress some of the previously described deficits in research that purports to examine the relationship between science and society. As an ecosystem science does not exist in isolation but is instead part of a larger social system. It is deeply influenced by the interests and discourses of nonscientists. Furthermore social ecosystems are created and maintained through both private and public activity. What appears to be individual actions are coordinated with other social actors. This shapes both the experience of science and the linguistic interpretation of that experience. Lastly, ecosystems are sustained through the development of functional, although not always equitable, relationships among their inhabitants. To survive and thrive, the system must achieve balance, stability, and openness to innovation. Thus boundaries are central to the functioning of any ecosystem. They give shape to the spaces they contain, relate one system to another, and mark the regions in which the social meaning of scientific practice is negotiated.

Scholars studying boundary work have suggested that one way rhetoric serves to define the substance and practice of science is by reinforcing the distinction between science and nonscience. This type of discourse is often characterized as the scientific community's response to outside challenges to its authority and autonomy, as in Taylor's account of refutations of creation science.[13] As the latest round of debates regarding "intelligent design" illustrate, demarcation issues also emerge in rhetoric that originates in the public sphere. Anne Holmquest has explored public construction of science in her study of media portrayals of science and the use of scientific evidence in the judicial system.[14] In both instances the influence of public discourses is readily apparent and the process of constructing boundaries appears to be a dialogue. However, focusing intently on the discursive nature of scientific boundary work should not tempt scholars to overlook the rhetoric of the material culture of science. Peter Galison and others have acknowledged the significance of material objects, the culture associated with them, and technology in scientific work, examining the sociological and rhetorical significance of boundary objects in scientific work.[15] Such work attends specifically to the articulation of regions between and among institutions and disciplines. While addressing the role of language, this work also explores the role of image and material culture in the coordination of scientific practices and the definition of scientific communities and professions.

Much of the existing work in science studies attempts to explain how communication demarcates science from other social institutions.[16] A number of studies assert that the definition of science begins with the articulation of specific norms and values that express scientific character.[17] Such rhetoric constructs boundaries in two ways. First, it often defines the standards of conduct or performance within a given scientific community. Second, and equally important, it distinguishes the behavior of scientists from that of nonscientists.[18] Thomas Gieryn defines this type of boundary work as discourse that allows the scientific community to fend off challenges to its autonomy and authority: "Boundary work is a rhetorical style in which scientists describe science for the public and its political authorities, sometimes hoping to enlarge the material and symbolic resources of scientists or to defend professional autonomy."[19] Gieryn emphasizes the adaptive nature of boundary work. He argues, "The boundaries of science are ambiguous, flexible, historically changing, contextually variable, internally inconsistent and sometimes disrupted."[20] Boundary work is particularly functional in that it gives scientific rhetors the flexibility to address multiple audiences, both inside and outside of the scientific community in question. When the scientific community separates its values and ideals from the larger culture, it serves two rhetorical goals. First, it reinforces a sense of identity and community among scientists. Second, it bolsters the general sense of separation between science and other groups. Separation and distinction serve to preserve scientific power and distinguish between scientific and competing, nonscientific interests. Boundary work is no simple matter, however. As Gieryn details: "Characteristics attributed to science are sometimes inconsistent with each other because of scientists' need to erect separate boundaries in response to challenges from different obstacles to their pursuit of authority and resources . . . the boundaries are sometimes contested by scientists with different professional ambitions . . . ambiguity results from the simultaneous pursuit of separate professional goals, each requiring the boundary to be built in different ways."[21] Flexible and adaptive, it is nonetheless complex as a means to deflect challenges from those who would threaten the identity and autonomy of the scientific community.

Holmquest extends the concept of boundary work to include public discourse and explores how demarcation functions in other spheres of argument. Holmquest strives to combine Gieryn's insights with Thomas Goodnight's notion of spheres of argument. Her primary concern is to demonstrate that the scientific community does not always affirm its expertise and autonomy as an exclusive source of knowledge. In cases where "people draw the line between the democratic principle and the profession," Holmquest demonstrates that boundary work is not always affirmative and disciplinary in nature and that it is sometimes controlled by those outside the scientific community.[22] Unlike Gieryn, whose work is focused solely upon the discourse of scientists, Holmquest asserts that our understanding of boundary work can be extended only through a close examination of the arguments offered about science by both scientists and nonscientists. We must examine the push and pull of competing definitions to understand how and why the boundaries of science articulate with the public sphere. As Holmquest writes, "By considering the details of [a] case in which a presumption of expertise was resisted by a science and 'courted' by the law, I have provided proof that the demarcation between science and non-science was decided by the way in which doubt was suspended, organized, and re-introduced into arguments."[23] By taking this approach, Holmquest argues convincingly for a pragmatic perspective on boundary work. Instead of viewing the process of demarcation as *inherently* defensive, as does Geiryn, she casts this type of rhetoric as an "opportunity to redirect the activities of human reason."[24] Holmquest's approach seems particularly useful when considering the rhetoric of Fermilab. As subsequent chapters will reveal, the nature and value of particle physics research long has been disputed. Those within and outside of science often have differing opinions about how, where, and by whom the future of the field should be decided.

As indicated earlier, Charles Alan Taylor takes the most expansive perspective on boundary work and its relationship to demarcation. He argues that demarcation is one among many types of rhetorical practice designed to define science. He classifies boundary work as an inherently public form of demarcation and argues that "Public demarcation rhetorics appear motivated by the specific goal of establishing the boundaries of science for audiences which do not participate in the day to day activities of science. Those who do participate in such activities, professional scientists, are not likely to explicitly concern themselves with defining what it *means* to do science."[25]

Working from this public locus of demarcation, Taylor hopes to expand our notion of demarcation to include the kind of tacit knowledge that permeates the everyday practice of science.[26] He argues that to understand fully the demarcation of science in society, we must explore how science is articulated in both discourse and in practice: "In rhetorically confronting the myriad technical exigencies of scientific practice, such as error correction, report writing, and the like, scientists *implicitly* construct and reconstruct what it means to do science. . . . Viewed in this way, demarcation is less a matter of strategy than a matter of ongoing practice."[27]

While Geiryn focuses on boundary work as a reactive or defensive rhetoric, Taylor is concerned instead with its constitutive nature. In particular, he deepens our understanding of "boundaries" by considering the rhetoric of scientific practice. Many philosophers and sociologists of science assert that science is defined in practice. If this is true, then demarcation rhetorics should be evident in both the talk and behavior of scientists. As he moves into the realm of scientific practice, Taylor reinterprets boundary work as an ongoing, rather than episodic, phenomenon. Behaving "like a scientist" has normative force that defines the very nature of science for those within and those outside a given scientific community. The substance of science must be articulated through such regular activity so that when faced with challenges from other interests, the community has sufficient rhetorical resources to argue for its uniqueness and constitute its autonomy.

The shift to practice is useful for the study of Fermilab since the changing practice of particle physics is both a function and an outgrowth of demarcation discourses. As the future of basic research has been increasingly negotiated in the public sphere, the practice of high-energy physics has changed.

The community has consolidated many of its research efforts over the course of the last thirty years, eliminating individual communities or modes of research and reshaping the nature and experience of scientific training. These choices, prompted in part by discourses outside the scientific community, have dramatically altered the internal boundaries of the field.

In sum demarcation rhetorics can be theorized in a number of ways. For Gieryn demarcation rhetorics serve multiple audiences to multiple ends. They are goal-directed, flexible, adaptive, and aimed largely at helping the scientific community protect its autonomy and identity from outside challenges. The flexibility and historicity of boundary work is equally important for Holmquest, and yet she emphasizes that these qualities should be regarded as evidence that boundaries are contested by other cultural institutions that can and do lay claim to scientific practice in the name of public interest. Taylor attempts to deepen both these interpretations, considering explicit and implicit manifestations of boundary work in scientific practices and the relationship of these practices to more general rhetorics of demarcation. Following Holmquest and others concerned with the practical consequences of rhetorical criticism of science, Taylor's work examines how demarcation rhetorics shape science and scientific practice on all levels, public and professional. I make use of each of these interpretations in my attempts to explain Fermilab's development and changing goals, the particularity of arguments designed to protect and defend the autonomous professional practice of high-energy physics, and, most important, the myriad of rhetorics—from practice and performance to image—implicated in the construction of laboratory's boundaries.

Demarcation rhetorics both manage and respond to the influence of *nonscientific discourses*. As a result they regulate the many influences on scientific communities, practices, and institutions. Yet most studies of demarcation rhetorics still remain science-centered, focusing on the discursive activities of scientists themselves despite many indications that these rhetorics themselves are prompted and shaped by outside discourses. Thus the mutual influence of scientific and nonscientific discourses on the development and management of boundaries has still been insufficiently explored in existing literature. In his expanded treatment of demarcation rhetorics, *Defining Science,* Taylor calls attention to this deficit in the literature. He suggests expanding not only our knowledge of the participants or constituencies in demarcation rhetorics but also the very nature and functioning of demarcation processes: "The cultural configuration that privileges science over and against politics (and other cultural discourses) is fundamentally discursive. To suggest that any hierarchical configuration of practices is a rhetorical construction entails as well that it is subject to alternative construction—a task to which rhetorical critics should set themselves."[28]

Simply recognizing that scientists and nonscientists alike participate in the demarcation of science is only one aspect of such a critique, however. We must also acknowledge that the origin and form of demarcation rhetorics vary widely among contexts. Demarcation discourses emerge as often through mundane activities as through specialized discourses. Whether we are or are not scientists, our fundamental understanding of science and its role in our lives is constituted as much through everyday life as through any formal expression of the abstract norms and values of science. Scientists may, from time to time, formally define and discuss the nature of science. Far more often, however, it is their day-to-day activities in the laboratory or the classroom that define their membership in the scientific community. This is arguably one of the most important observations made in *Defining Science*—the recognition that boundary work and demarcation are central to the practice of science, the development of scientific research programs, and the development of science policy and vice versa. "Demarcation, then, is not a phenomenon (rhetorical or otherwise) which is somehow divorced from the conduct of scientific activity. Indeed, it is a central (constitutive) constraint on the everyday productive and valuative practices of scientists as laborers. The rhetorical process of demarcation is part of the 'work of science.' The principles and norms which have traditionally been said to define science are meaningful only in social action, in praxis."[29]

Thus boundaries are the product of experience of and for scientists and nonscientists alike. Scientists derive identity, meaning, and a sense of separateness from what they do and do not do. But much more is at stake than simply the construction of difference. Rhetorics of difference constitute categories, but only boundary work negotatiates the relationships among these categories. In doing so boundary work defines not only the nature of science, but also its social worth and in consequence the very policies that govern the practice of research.

As Fermilab articulates its relationships to its various stakeholders, it necessarily defines and redefines itself. Any attempt to understand the communication between the laboratory and its publics must therefore account for the ways that discursive boundaries allow scientists to define and maintain unique discourse communities that most often evolve in opposition or contrast to other social groups. Over the course of Fermilab's history, for example, the institution has negotiated a number of relationships with those living outside its physical boundaries. At times, for example, the lab presents itself as a resource for those who live in the area. Neighbors are encouraged to take advantage of the grounds as a place to exercise or enjoy nature. In doing so it includes both the neighborhood and the neighbors. The lab has also presented itself as an "educational resource," a designation that may or may not honor the boundaries of the neighborhood. FNAL also frequently represents its work as an investment that will return technological benefits to the American taxpayer, a metaphor that stretches the boundaries of the public and the laboratory further still. Of late the laboratory has met with a new challenge—convincing neighbors that the lab and its work poses no health threat in the wake of on-site detection of tritium. Each of these representations of the laboratory is embedded within a different relational context. Individuals relate to their neighborhood nature park differently than they do to their investments. The degree to which the general public is willing to challenge or question the work of the lab is determined in part by the nature and form of the relationship between science and society constructed through boundary discourses. In this sense the location and perceived permeability of boundaries separating various publics from the laboratory constitute relationships that in turn influence behaviors both inside and outside the laboratory.

Visibility and Priestly Rhetorics

The reality of scarce resources and a history of strained relationships between big physics and its publics demand that scholars should account for the boundaries that define current cultural understandings of the relationship between science and society. With its funding maintained in large part through public discourse, the rhetoric of high-energy physicists must do more than simply preserve the identity and autonomy of the community. It must also justify high-energy research to the public and encourage the continued support and funding. The articulation of this complicated persuasive task is an understudied outgrowth of boundary work. The degree to which an individual scientist can travel back and forth across the boundaries and bridge the rhetorically constructed divide between science and society is of particular importance.

A number of studies along these lines have suggested that the divide between science and society is negotiated through the identity and actions of individual scientists, most notably "stars" who represent the community in the public sphere. These visible scientists act as mediators between the scientific community and the general public.[30] Rae Goodell's work on "the visible scientist," for example, explores the social capital of popular scientists such as Linus Pauling and Carl Sagan. Goodell's work raises a number of questions relevant to boundary negotiation, including the degree to which, once they become visible and celebrated in popular culture, visible scientists can remain viable members of their respective scientific communities. While scientists must always negotiate the conflicting norms of communalism and individual achievement, Goodell argues that most scientific communities will eventually marginalize colleagues who become too visible in nonscientific realms. To distance the community from a visible colleague is not without risk, however. Visibility can reap significant benefits for both the individual scientist *and* the community they represent. Carl Sagan was frequently marginalized by other scientists for his work with the Search for Extraterrestrial Intelligence (SETI) program, but his visibility on this issue (and its subsequent uptake in popular culture) undoubtedly increased public awareness of radio astronomy.

Not surprisingly Sagan's stardom has also prompted close examination by rhetoricians and other science-studies scholars. Michael Mulkay has studied how stars within a scientific community are used as representatives of the community, not as real scientists but rather as spokespersons for projects or interests that require the attention or support of outside communities.[31] Thomas M. Lessl has described how Sagan's identity was affected as a result of this visibility. Sagan was neither fully scientist nor fully "social." He was instead the very expression of the boundary between science and nonscience—the division between science and society was articulated in and through his very existence. In this role he exemplified science through his enacted or performed separation from the world of mundane concerns and his focus on higher, other worldly goals. And yet, as a visible scientist, he traversed the landscapes of popular media in order to model the "proper" relationship between an individual and science or scientific modes of thought. Such discourse produces a gradual and complementary humanization of science and scientization of the public. Over time, as Lessl and others argue, these rhetorics may strengthen public support of science even as they marginalize the individual visible scientist.[32]

Studies of this sort have particular relevance for the analysis of communication at Fermilab, as visible scientists have strongly shaped its discourse. As representatives of the laboratory community, visible scientists embody the institution and its values for various publics. From its earliest origins in the Manhattan project, the national laboratory system has been a pathway to and from visibility. For example, Robert Oppenheimer was both celebrated and vilified as a consequence of his leadership at Los Alamos. As "father" of the bomb, Oppenheimer became a symbol of society's hope, fear, and anxieties about the future of physics. He was simultaneously heralded and scapegoated for his complex relationship to the technologies and institutions of wartime physics. In more recent eras, national laboratory directors have embodied the political power of both the science and the system that sustains it. Wolfgang "Pief" Panofsky, the first director of Stanford Linear Accelerator, or SLAC, was known for his tenure as Eisenhower's science adviser and his subsequent work as an expert witness on the development of nuclear technologies. John

Marburger, recent director of Brookhaven National Laboratory, graduated to a post in the George W. Bush White House. Each of these individuals was visible not only in their organizational role but also as experts who weighed in on critical debates that define the boundary between science and society at a given time. For example, Oppenheimer and Panofsky have each publicly cautioned against the proliferation of nuclear weapons.[33] Leon Lederman, FNAL's second director, has been a vocal and prolific advocate for changes in science education. In the context of a public debate about nuclear weapons or inequities in the education system, these scientists do more than offer expert testimony—they embody scientific authority and define civic responsibility as a function of science.

In the hierarchical environment of particle physics, the laboratory director functions as the symbolic embodiment of the institution and the science it represents. He or she enjoys a great deal of power with respect to decisions regarding the future of the laboratory and sets the tone for both the practice and representation of the lab's work. Directors are expected to have a vision for the institution, to lead according to that vision, and in many ways to become that vision. As priestly rhetors, they embody the authority of their institutions for both the general public and government officials who make funding decisions. As Taylor argues, their primary function "is the clarification [construction and reconstruction] for nonexpert audiences of the meaning of science."[34] In the case of Fermilab, the meaning of science is synonymous with "basic research." From the perspective of the laboratory director, the field of high-energy physics will only survive if the public accepts that research is intrinsically worthwhile and need not be linked to pragmatic technological outcomes.

Detailed analysis of what Thomas Lessl refers to as the "priestly" rhetoric of many publicly visible scientists can add further depth to the previously described theoretical frameworks.[35] When visible scientists serve as "priests," they create even greater power for scientific communities by constructing social relationships that privilege scientists over and against other members of society. In order to secure public support for science, priestly rhetorics deploy ideological arguments that secure and reinforce power relationships between science and the larger culture. The success of such an approach is determined by its capacity to place science in a hierarchical relationship to other pressing public concerns or human values. Visible scientists work to embody, reframe, and reinterpret the relationship between science and society in an effort to maintain public support for research. Priests transform the character of the scientific institutions for their audiences. Rather than separate science from society through a defensive rhetorical strategy, priestly rhetorics often function in positive terms. Priests are exemplars of human scientific behavior that demonstrate the difference between scientists and nonscientists and, more subtly, suggest the superiority of the scientific worldview in a given case. As Taylor suggests, "The definitions of appropriate scientific practice which are rhetorically negotiated by practitioners quite often form the content of public accounts of those practices."[36] And yet definitions formulated in the laboratory can function quite differently when deployed in the public sphere. Priestly rhetorics often serve to reinforce scientific authority in the larger culture, particularly when that authority may be under threat. As embodied by the visible scientist-priest, science is not merely one among many admirable human endeavors, but rather the pinnacle of human achievement. Similarly, basic research is often elevated to a priestly, almost monastic, pursuit. Once characterized as the pure expression of science, it is more easily defined as worthy of public support. In short, priestly rhetorics suggest that humanity is at its best when it expresses the values and practices of science.

The works described here provide a critical foundation for understanding communication between scientific and nonscientific publics. A rhetorical approach to the study of boundary work can help to answer important questions about the maintenance of identity, power, and influence in the relationship between science and society. The approach encourages scholars to look beyond dominant interpretations of scientific communication as the value-free dissemination of knowledge and reframe the construction of science as a rhetorical act shaped by culture and experience. While much of the described work acknowledges that science must communicate with lay publics to facilitate its work, research to date has focused almost exclusively on those within or strongly connected to science. If work in science studies is to fulfill its promise of a more complete understanding of the relationship between science and society, more scholars must attempt to account for the experience and discourse of science among nonscientists. Study of the situation at Fermilab offers an example of one possible approach by focusing on how science and technology are made meaningful within a larger social context.

Rhetorical Cartography

To date, most explorations of boundary work have been based in traditional methods of rhetorical analysis, most commonly textual analysis. Yet the social construction of scientific institutions is a complex process involving many layers of meaning and interpretation encompassing texts and performances and a wide spectrum of verbal and visual rhetorics. While the boundaries of scientific practice and knowledge are made visible in the texts explored in the previously mentioned studies, the experience of those texts is largely inaccessible through traditional methods of rhetorical analysis. By combining an analysis of written and visual texts with interviews and observation, this project attempts to fill that gap. It examines the relationship between varied experiences of science and the discourses that emerge from those experiences.

This combined approach offers several advantages. First, it provides a means to capture and assess the formative contexts of boundary rhetorics. This allows us to understand better and interpret how both texts and experiences enter into deliberations about the nature and social worth of publicly funded scientific projects. Second, observation and interviewing provide more direct access to scientific practice. While analyses of textual or performative representations of scientific practice are critical to understanding boundary work, they should be supplemented and expanded through the observation of scientific practice and discourse inside a given organization or laboratory. Third, interviewing and observation are particularly effective ways to access the public's discourse about science. Unlike that of the scientific community, whose organizational structure fosters collective rhetorical activity, the public's voice tends to be fragmented. Interviews and observation can capture and account for some of the many ways in which the laboratory is understood by those outside its gates. Last, a multiple methodological approach is the only way to account for the interactions between texts and experiences. Only by asking questions and observing responses can we assess the impact of the existing boundary rhetorics. While textual analysis provides a detailed understanding of the discourse invoked by those within and outside a given scientific community, interviewing and observation allow us to see and assess the effect of those discourses in real-world science/society relationships.

Fermilab as Text and Image: Visual Rhetoric

Robert Wilson's purposeful effort to create a "scientific utopia" suggests the importance and necessity of attending to both the verbal and visual dimensions of the laboratory environment. This study considers the laboratory's physical environment, public displays, and written materials as texts strategically created to communicate the identity and values of the FNAL research community as well as to manage the lab's relationships with the publics it serves.

Defining the lab's textual dimensions is a complex process, one that demands the critic to attend simultaneously to the undeniable influence of individual scientist-rhetors as well as the lab's placement within a larger cultural milieu. The lab's textuality operates on at least three different levels. Individual verbal and visual textual elements—architecture, sculpture, exhibits—have meanings that emerge from both their creation and their intended rhetorical objectives. The meaning of these elements within the FNAL research community changes over time as individuals and groups find new ways to relate to the lab's physical spaces, technology, and research programs. For those who are not members of the scientific community, meanings are shaped by changing relationships between scientific institutions and the larger culture. To account for these multiple layers of meaning and interpretation, the analysis presented in the following chapters employs both historical and critical-textual perspectives to understand better how the lab's meaning has evolved over time and across communities.

Several additional important theoretical and methodological issues merit consideration. The lab's rhetoric is largely visual in nature, and the chosen method of analysis must reflect this fact. The visual orientation of FNAL rhetoric derives from three sources: the strong visual orientation of contemporary high-energy physics as a field; a concerted effort to simplify mathematical concepts for nonscientist audiences through the use of visual metaphors; and the artistic and architectural contributions of the lab's first director, Robert Wilson, and its in-house artist, Angela Gonzalez. The visual is most often used to stand in for or point to meanings that are difficult to express verbally. Often a more diffuse aesthetic experience that replaces the precision of one of the community's primary symbol systems, mathematics, results.

To understand fully the significance of image and display at FNAL, we must ask, "How are the boundaries between the laboratory and its publics constructed and enforced through aesthetic experience?" Geiryn's use of the word "style" to describe boundary work gives us occasion to consider the persuasive dimensions of aesthetic experience and its relationship to boundary work. As Robert Hariman argues in his work on the political style: "Style becomes an analytical category for understanding a social reality; in order to understand the social reality of politics, we can consider how political action involves acting according to a particular *political style*. From this perspective, political events are produced within a social setting through conventions of artistic composition depending upon aesthetic reactions for persuasive effect."[37]

Comparisons can be drawn between Hariman's analysis of political style and the rhetoric of science where the performance of community identity is strongly linked to social power. In the realm of science, too, style communicates as forcefully as argument. Style has the power to generate meaning and experience not accessible through verbal communication. FNAL draws heavily on the unique image-culture of high-energy physics to communicate the nature and worth of its work and the ethos of the experimenter. Accordingly this book draws on methodological models that confront the visual and aesthetic directly rather than trying to force the elements and functioning of visual texts into verbal argumentative categories. In doing so my analysis recognizes that style is as much enacted, displayed, and performed as "read."

Analyses of visual rhetoric often highlight important issues of textuality and authorship, these issues are critical to understanding the evolution of FNAL's rhetoric. For example, in their article on the Vietnam Veterans' Memorial, Blair, Jeppeson, and Pucci discuss how the visual elements of the memorial make apparent the distinctions between work and text: "The distinction between treating rhetoric as a work and as a text becomes tremendously important.... Postmodern rhetoric recommends a critical practice that considers pregiven material unity irrelevant. To treat a postmodern discourse as a complete or unified structure is to utterly miss the point. The goal of a textual reading is to grasp the multiplicity of any discourse; to constrict or expand the scope of a text is to yield very different readings."[38]

In reading the memorial as multifaceted "text," Blair, Jeppeson, and Pucci are led to question its authorship and symbolic unity. Is the meaning of the memorial contained in the architects' vision of the monument, in the visitor's experience of it, or in the critic's analysis? FNAL presents similar questions with regard to the distinction between work and text. To be sure, much of the lab's rhetoric, including especially its stylistic-aesthetic rhetoric, can be attributed to Robert Wilson and the specific ways in which his vision of scientific utopia are inscribed in the spaces, buildings, and sculptures he designed for the laboratory. To focus exclusively on the aesthetic of Wilson's vision, however, is not only to lose sight of the political intentions and consequences of his efforts but also to close interpretive access to a myriad of other possible readings that might emerge from one's experience of the laboratory. This study attempts to "grasp the multiplicity" of FNAL's discourse by examining multiple critical-textual perspectives.

Taking this approach highlights the political nature of the lab's boundary discourses. Most studies of boundary work acknowledge that conflict between scientific-technical and other social groups often initiates episodes of boundary construction or reinforcement. These studies sometimes fail to grasp the consequences or implications of these encounters for the relationship between science and society. Every instance of boundary-constituting rhetoric is in effect an episode of political discourse. Initiated by challenges to the status quo, boundary work either reiterates or redefines a relationship between scientists and the challenger group. Thus boundary discourses are used not only to reinforce the identity of those inside the boundaries but also to silence those outside the boundaries. To acknowledge and theorize further the political nature of boundary work, this analysis is guided by a concern for how FNAL's rhetoric "exerts force ... within its field of action."[39] Fermilab's boundary work affects not only how scientists see themselves and their work but also relationships with outside groups that either support or restrict the continued operation of the laboratory.

Fermilab as Narrated Experience

Though the current rhetorical literature offers no prototypes for the study of an individual's lived experience of science, interpretive analyses of communication provide ample theoretical and methodological models that can be used as starting points. At the intersection between argumentative and aesthetic rhetorics lies narrative. Among the most productive and insightful works about this point of intersection are studies generated from narratological and social constructivist perspectives. These studies focus on the ways in which discourse constitutes, shapes, and reflects experience in stories and myths.

Narrative theory describes how complex webs of representation coalesce into coherent stories of everyday life. Common events are brought to life through the act of telling one experience to others. The narrative produced and reproduced through communication organizes experience for both listener and narrator.[40] This perspective aids us in understanding the meaning of science in daily life. The most salient of our scientific experiences are grounded in a stockpile of narratives that explain our relationship with science. Working scientists, for example, will draw upon different narratives than those who consider themselves laypersons. In order to support and maintain our existing frameworks of meaning, new narratives tend to resemble and reinforce old ones in both form and content.

Over time individual stories about science experience crystallize into coherent social frameworks that shape the ways in which such tales are told in a given culture. The stability of an individual story is necessarily limited by both the nature of the experience and the cultural horizon that shapes the interpretation of that experience. Schutz described the mutual influence of language and culture as socially originated interpretive constructs that he referred to as stocks of knowledge: "These images, theories, ideas, values, and attitudes are applied to aspects of experience, making them meaningful. Stocks of knowledge are resources with which persons interpret experience, grasp the intentions and motivations of others, achieve intersubjective understandings, and coordinate actions."[41]

While stocks of knowledge of this sort are not in and of themselves narratives, they do have important consequences for both the formation and interpretation of narratives. As resources through which we interpret the world, stocks of knowledge are not neutral commodities. They give a definitive texture to an individual's relationship with science by guiding our interpretation of experience. For example, a tritium leak at Brookhaven National Laboratory (BNL) was clearly experienced differently by those who worked there and those who lived in the surrounding community. The different stock of knowledge on which these two groups of individuals drew not only structured their opinion of the laboratory but also affected their reception of arguments supporting increased DOE research funding. Put simply, different interpretations resulted from very different "rhetorics of everyday life" in and around a research community.[42] These different communities were telling different stories about radioactivity and accidents. As this example illustrates, narratives constitute far more than an end in themselves. They are also a resource for the researcher in that they can reveal the political dimensions of seemingly nonpolitical experience. Analysis of narratives generated in response to a lab visit can reveal a participant's views about FNAL and also uncover interesting and unexpected opinions about the social worth of basic physics research. When considering how the nonscientist confronts the abstract and frequently paradoxical world of high-energy physics, it is helpful to remember that "in all our knowledge of ourselves and in all knowledge of the world, we are always already encompassed by the language which is our own."[43] Acknowledging the complex interplay between discourse and personal experience is the first step toward a more accurate understanding of the relationship between science and society.

Unfortunately traditional modes of research in science studies are often blind to the interplay between discourse and experience. Although observation has been used widely in the narratively mediated study of working scientists, this methodology offers few advantages for studying nonscientists whose encounters with science are embedded in the complex tapestry of daily life and thus hidden from public view.[44] Furthermore observation is simply not sufficient to discern and make sense of a nonscientist's experience of science if we fail to account for communication and interpretation. Only interaction with participants on both sides of the divide between science and society will reveal what each considers to be science and its meaning.

Similarly survey methodologies, although excellent vehicles for testing knowledge or assessing opinion, are ill-equipped to capture the diverse nature of lived, narrated experience. Understood as meaning produced in situ, experience is not easily reproduced through survey questions designed to fix scientific knowledge outside of the context of its production. As with observation, survey answers can tell researchers what a respondent knows or thinks, but they provide little information as too *how* or *why* they produced that particular response. Further interaction is necessary to link survey answers to actual experience—and that further interaction will necessarily turn up tacitly structuring narratives.

Neither abstraction nor surveys generate sufficient discourse to capture the contextual and often idiosyncratic meaning of science in and for an individual life. Furthermore, because these methods produce discourse-poor data, subsequent interpretation often reproduces the dominant interpretation of the relationship between science and society. Such readings may actually obscure the contextual, embedded, contingent nature of the nonscientists' experience of science. Thus in order to achieve my stated goal of demonstrating how rhetoric shapes the development of the relationship between science and society, this project has incorporated multiple methodologies capable of producing discourse that demonstrates the varied and sometimes surprising ways in which individuals make science meaningful in their everyday life.

Interviewing, and with it the production of conversation, offers a viable and underutilized alternative to other methodologies. Active interviewing is designed to confront both meaning production and meaning produced. The interviewers strive to tap into discursive processes hidden from observational and survey methodologies in order to provide an opportunity to better understand the "how" behind the "what" of an observed behavior or survey answer: "understanding *how* the meaning-making process unfolds in the interview is as critical as apprehending *what* is substantively asked and conveyed. The *hows*, of course, refer to the interactional, narrative procedures of knowledge production, not merely to interview techniques. The *whats* pertain to the issues guiding the interview, the content of questions, and the substantive information communicated by the respondent."[45]

The interviews on which this book draws can be viewed as purposeful conversations. The interviews generate content in the form of answers and opinions converging on assumed stories. They also reveal linguistic resources that both the respondent and the interviewer have used to make sense of the topic.[46] Most important, active interviews generated narratives about the many ways in which nonscientists make sense of and use science in their day-to-day existence. Subsequent closer analysis reveals the frameworks or stocks of knowledge that shape participants' stories and thus enrich understanding of their experiences of science. Interviewing was most beneficial when it was used to engage respondents in a focused, but reflective, conversation about science that explored any and all ways in which science is meaningful in their lives. For example, the scientists interviewed were asked to reflect on how they explain their work to the public. Visitors were similarly asked to reflect upon their experience of the self-tour and the ways in which they related their visit to the laboratory to their everyday life. Both laboratory employees and visitors were encouraged to tell stories that reflected their experiences of science. In this sense the interview provided a unique opportunity to theorize how texts constitute audiences and how audiences interpret texts.

Seen from this perspective, the interview functioned as both a methodology and an object of analysis. As a methodology, interviewing balanced structure against spontaneity. Discussions were focused, but lines of inquiry could be modified to suit the unique experiences of individual respondents.

Interview guides developed several specific lines of questioning regarding respondents' experiences at the lab and their views about the role of science in their lives and in society as whole. In this sense the "interview guide" promoted flexibility and created space to accommodate an individual respondent's interests. Though interview guides provided the basic framework for questioning, interviews most often developed into a dialogue between the researcher and the respondent. Thus the validity of the results detailed here should be judged not by rigid adherence to a preordained questioning procedure, but by the degree to which the interview reflected the experience of the participants and their shared negotiation of key texts.

Once produced, the interview transcripts that I generated in the course of my fieldwork became objects of analysis that could be viewed from both diachronic and synchronic perspectives. Researchers all too often restrict dialogue within the discrete boundaries of the research encounter, failing to view the interview within its larger sociohistorical contest. Focusing on the play between experience and meaning, I compensated for this defect by regarding the interview not as an isolated encounter but as one episode in the ongoing conversation of a participant's life. Although the interview was transformed eventually into a text that encapsulated an individual's relationship to science, the participant's life provides an ever-changing backdrop that challenges each new reading of that text. Interpretation and understanding are thus informed by the knowledge that "text and context are in a continual state of tension, each defining and redefining the other, saying and doing things differently over time."[47] In the case of this project, textual interpretation was informed both by the respondent's role with respect to FNAL (employee, visitor, and so on) and also by their ongoing membership in a scientifically and technologically advanced culture.

The combined use of rhetorical-textual analysis and interviewing has revealed a number of interesting findings. Most notable among these, as we will see, is the laboratory's recurring rhetoric of the technological sublime and the public's problematic, often-puzzled response to this rhetoric. Accordingly the following chapter details the visual rhetoric and institutional narratives that structure the experiences of lab employees and visitors. I examine the nature, origin, and evolution of the lab's interpretation of its identity through the lens of the technological sublime. I also consider the influence and role of the laboratory successive directors in the development and articulation of this rhetoric.[48]

⟩⟩ THREE ⟨⟨

Rhetoric, Persuasion, and the Sublime Laboratory

Fermilab appears simultaneously extravagant and understated, excessively beautiful for a scientific institution and markedly self-effacing about its own grandeur. As you are drawn into the site, you see first the beauty of the natural world—prairies, stands of oak trees, and ponds alive with migrating birds. These comforting images are soon sharply countered by the challenging extremes of technology—capacitors and power poles that somehow seem to manage unimaginable volumes of electricity; dozens of helium tanks to keep the accelerator cool; a dramatic orange and blue "box-building" that houses one of the lab's two massive particle detectors. The juxtaposition of technological power and natural beauty is in itself sufficient to prompt a powerful response, and yet it is the hidden presence of the accelerator itself that organizes our experience of the lab. As you enter, you look for it. You know it is out there somewhere, but it is surprisingly hard to find. Eventually you may notice it resting quietly under an earthen berm immediately to the east of Wilson Hall, the lab's main building. The perspective from which we view the accelerator is key to its power and role within the lab's rhetoric. As we drive by, we can glimpse at best only a portion of the machine. From the fifteenth floor of Wilson Hall, however, we are positioned squarely within the laboratory's official public narrative and come to "see" the accelerator in an even more dramatic light.

We have come to regard aesthetic experiences as a poor cousin to rational persuasion, particularly when it comes to science. Beauty and emotion seem somehow antithetical to the objectivity and dispassion normally associated with science. Considered more carefully, however, the emotional tensions provoked by visiting Fermilab exemplify its unique rhetoric, a powerful mix of rationality and aesthetic experience. This chapter considers the significant elements that contribute to this unique experience, in particular the use of visual, experiential, and performative rhetoric to shape audience perception of the laboratory. The first sections consider how the laboratory's rhetoric is shaped by the American technological sublime. The remaining sections consider how the rhetorical performances of the laboratory's first two directors,

Robert Wilson and Leon Lederman, contributed to public perceptions of the laboratory and its work.

The American Technological Sublime at FNAL

Sublime rhetoric structures our experience in terms of the dividing lines between approach and avoidance, awe and fear, and action and inaction. Hariman noted these oppositions at work in political style and offered the following explanation of the role of "aesthetic knowledge" in rhetoric: "The sublime refers to that preeminently aesthetic sense of wonder, expansiveness, and awe that we experience in the face of natural beauty, technological power, or artistic perfection. The sublime is created when some limit condition in ordinary consciousness is exceeded, often because of some assertion or confluence of forces that catch us by surprise. . . . In the sublime there is the paradoxical simultaneity of seeing beauty and experiencing power: we see an aesthetic object, separate ourselves from ourselves because [it is] so beautiful, and feel an enormous transfer of energy that sweeps us into a transformed world."[1]

The rhetoric of Fermilab is an instance of the sublime. As Hariman suggests, the visual grandeur of the laboratory exists in inextricable interplay with the mundane discourse of the management, funding, and application of scientific research.[2] The laboratory's sublime construction of its work and its technologies thus simultaneously references and refutes traditional, rational justifications for government-funded science. Visitors, perhaps initially motivated by curiosity about how their tax dollars are being spent, are invited into the extraordinary transcendent world of particle physics where questions of "why" and "how much" suddenly seem odd and even irrelevant. The sublime experience of the lab's technology in particular—functional and yet undeniably awesome in both scope and power—directs attention away from seemingly obvious realms of pragmatic inquiry and critique. But at the same time visitors are prompted to support the laboratory's world with their tax-paying pocketbooks. It is a unique and at times quite effective way to manage the laboratory's rhetorical boundaries.

To understand better how the Fermilab's rhetoric functions to manage boundaries and relationships with various publics, it is helpful to examine the roots of the technological sublime in American culture. David Nye defines the American technological sublime as "an essentially religious feeling, aroused by the confrontation with impressive objects, such as Niagra Falls, the Grand Canyon, the New York skyline, the Golden Gate Bridge, or the earth-shaking launch of a space shuttle. The technological sublime is an integral part of contemporary consciousness, and its emergence and exfoliation into several distinct forms during the past two centuries is inscribed within public life. In a physical world that is increasingly desacralized, the sublime represents a way to reinvest the landscape and the works of men with transcendent significance."[3]

Nye charts the development of this peculiarly American formulation of the sublime by comparing and contrasting it with the features of the natural sublime theorized by Longinus, which influenced Edmund Burke and the philosopher Immanuel Kant.[4] Like the natural sublime, the technological sublime has the power to unite. Nye argues that the technological sublime "can weld society together," so that "human beings temporarily disregard divisions among elements of the community" in the force of something "bigger than themselves."[5] Such power does not signal exactly the private emotional response theorized by Burke or Kant. It is instead a social construction eliciting a social experience finely tuned to bridge societal divisions. In this sense, especially, sublime experience can be seen as, and actually is, rhetorical. It has the power to constitute publics in and around key objects or issues. Nye makes this point in the following way: "In the United States, where the sublime has increasingly become a group experience rather than a moment of private contemplation, these experiences often have overt political consequences, both as matters of public display and as issues of social policy. The questions of central concern in this study are these: *What objects have Americans invested with sublimity? What responses have there been to these difference objects? What is the larger ritual or political framework within which the sublime appears?*"[6] (emphasis added). The physical and social construction of the sublime at FNAL reflects this desire to unite a culture divided in its opinions and support for publicly funded research. It does not do so by accident. Disparate groups, both inside and outside of science, could be united through the shared, public understanding of the laboratory and its work when framed in these terms—and its promoters know it.

The origins of the strategy can be found deep in the lab's history. At the time of its founding, supporters sought to unite the scientific community and the government behind the symbolism of the accelerator and its research capabilities. The integration of these interests demanded the creation of new boundaries and new relationships within the realm of federally funded research. Particle physics desired to once again take center stage in the government-funded research program so as to garner support for the developing subdiscipline of high-energy physics. The prominence accorded to the field and its principal technology—particle accelerators—gave some in the physics community a platform from which to argue for a distinction between "basic research" and "applied research," a strategy key to articulating new boundaries and relationships between the Department of Defense and other federal funding agencies. In time high-energy physics was rhetorically cordoned off from weapons work even though it shared in essentially the same system of management and funding as other multipurpose laboratories that engaged in both basic and applied science. Thus unity was achieved, quite ironically, through division. Two distinct types of laboratories eventually emerged, defined by differences in technology, practice and, most important, orientation toward classified defense research. Those in the high-energy camp got the machines they desired, claiming all the while that their physics had no bearing whatsoever on or relationship to the physics being done in the so-called weapons labs. Meanwhile growth in the weapons complex continued almost unabated at places such as Los Alamos, Oak Ridge, and Livermore.

Burke and Kant had noted that the experience of the sublime presupposes but transcends fear. This distinction marks the difference between research and weapons labs. Whereas the natural sublime imbues ordinary objects with power by associating them with the sacred, the technological sublime creates power through the association of people with powerful technologies. These objects often have no reference point in the realm of the sacred but derive their rhetorical power instead from America's romance with technology. As in the Victorian era, when "new railway stations, aqueducts, factories, and warehouses were rhetorical structures, demonstrating the power of the builders," accelerators and weapons alike evoked a peculiar combination of powerful emotions.[7] Attraction, repulsion, terror, and astonishment have defined our cultural experience of nuclear technologies since the middle part of the twentieth century. These emotions also mark as exceptional those capable of creating such objects.

Nye traces this shift from the natural to the technological sublime to a time when the term "sublime" was also understood as a verb that meant "to sublimate"—not in the Freudian sense that it now bears, but in a sense still occasionally found in chemistry: "to act on a substance so as to produce a refined product," a sublimate.[8] "Alchemists seeking to bring substances to higher states of perfection employed sublimation in their efforts to attain the philosopher's stone. Alchemy gave the term 'sublime' a special coloring that anticipated the later response to industrial objects."[9]

As the production of sublime objects shifted from the divine to the human realm, so too did the power associated with them. The process of sublimation preserved the important relationship between nature and technology that aided our understanding of a human-created object as sublime, but it placed the purposeful actions of individuals at the center of the discourse. Whereas the natural sublime pointed to nature as evidence of the sacred, the technological sublime invokes nature only in order to highlight the human power to transform it.

The rhetoric of high-energy physics in the mid–twentieth century is not reducible to technology. It builds on the natural sublime by relying on a related discourse of sublimation that has been made particularly mysterious by ongoing revelations in the laws of quantum mechanics. On one level high-energy physicists make literal claims about their capacity to "refine" actual particles with accelerators and colliders. On another level they speak also of "refining" knowledge and understanding of the subatomic world. This "alchemic" discourse is in fact evident throughout FNAL's public texts. They include references, for example, that compare the lab to a "scientific crucible." The sophisticated linkage of the technology and its creators to the mysterious world of the subatomic suggests that, although visitors or funders can participate in the sublime experience of viewing the accelerator, the presence and authority of scientists are required to bring forth and manage the process of sublimation so that it remains sublime, not merely dangerous and fearful. Thus the technological sublime constructs authority through mystification and in doing so attempts to manage the flow of power between high-energy physics and its various publics.

These strategies resonate strongly with the rhetoric that defined early work on the atomic bomb—that of powerful men laboring to uncover the secrets of nature and harness its power. Regardless of the actual use to which their machines are put, physicists working on "peaceful" and defense-related projects participate in the same cultural rhetoric, a rhetoric grounded in America's unique role in the development of atomic technologies. Self-identified as "accelerator builders," Fermilab physicists emphasized their role in the design, construction, and control of a powerful new technology. The accelerator is regarded as more than a research tool for a knowledge-based community of scientists. It has also figured as a monument to the skill and creativity of its designers. In this sense accelerator builders in the 1960s and 1970s claimed an authority strangely similar to that once enjoyed by those associated with the Manhattan project.

In time, however, as the cold war grew more intense, the desire for increased funding and autonomy for basic research led many in the high-energy community to draw a sharp dividing line between research instruments and weapons. They commonly defined accelerators as productive rather than destructive. In the words of Sheldon Glashow, they were the "opposite of the bomb—Instead of using particles to produce energy, it uses energy to produce particles."[10] The laboratories that housed these machines were to be similarly distinguished—they were not weapons labs.

From this starting point we will examine first the articulation of a relationship between nature and accelerator technology through a narrative of sublimation. This narrative points inevitably to the role and importance of the accelerator's builders and users. Thus I go on to explore the unique role of laboratory directors, beginning with Robert Wilson, in the development and furtherance of this aspect of the laboratory's rhetoric. Of particular importance is the way in which the narrative of sublimation and the role of the machine-builder-user is reinforced by the experimental practice of high-energy particle physics in the United States and the redefinition of high-energy and accelerator physics as unique subspecialties within the larger realm of physics research.[11]

The Nexus of Nature and Technology: A Narrative of Sublimation

The central metaphors and analogies in the lab's rhetoric operate within a larger narrative of sublimation in which the accelerator becomes a tool for transforming both particles and people. Earlier I showed that the work being done in and by the particle accelerator is more frequently described in language suggestive of alchemy and sublimation than in terms traditionally associated with physics. Consider the following description of the Tevatron from a Fermilab informational brochure: "The Tevatron accelerates protons and antiprotons in a giant underground ring. When proton and antiproton collide at close to the speed of light, they make a tiny fireball of pure energy as intense as that at the big bang, when the universe was a trillionth of a second old. Some of the energy turns into matter, . . . yielding sprays of particles that may hold answers to our questions about the laws and origin of the universe."[12]

In this passage the capabilities and actions of the accelerator are reinterpreted through the technological sublime as giving access to the universe at the moment of its birth. The Tevatron is transformed into a "discovery machine." It simultaneously produces and discovers; thus it seems to take the place of nature. It creates what is already there and inaccessible to human beings. In this respect it is like the chemist's crucible, bound to nature and yet oddly superior to it. It produces both particles and knowledge. The image of sublimation provides additional touchstones for the interpretation of the accelerator. First, it defines the accelerator's purpose. In this frame the machine is designed to produce the most basic constituents of matter (and antimatter). The collision of protons and antiprotons is reinterpreted as the final step in a series of procedures that refine matter down to its fundamental state. As the accelerator facilitates the purification of matter, it also expands our understanding of nature and drives the development of new theories. In doing so, high-energy physics works toward an understanding of the fundamental nature of matter and reactivates conditions that have not existed since the beginning of the universe. Defining the accelerator in these terms—as a machine that somehow purifies matter and brings forth otherwise inaccessible states of material existence—elevates both the status of the scientific enterprise and the technology with which it works.

The act of colliding particles is both destructive and productive, simultaneously an act of annihilation (of protons and antiprotons in the case of Fermilab's Tevatron) and creation (of quarks and leptons). This apparent paradox of simultaneous creation and destruction is resolved through the transformation of matter into other forms. The laboratory's rhetoric places particular emphasis on the partnership of technology and skilled human operators in this process. Sublimation is key to the development of the technological sublime. As Eliade notes in his work on the origins of alchemy, "What the smelter, smith, and alchemist have in common is that all three lay claim to a particular magico-religious experience in their relations with matter; . . . All three work on Matter which they hold to be at once alive and sacred, and in their labours they pursue the transformation of matter, its perfection and its transmutation. This ritualistic attitude towards matter involves, in one form or another, man's intervention in the temporal rhythm peculiar to 'living' substances."[13]

The contemporary articulation of the technological sublime exists well removed from the realm of the sacred, even if "living," substance, and yet it sets up a comparable "intervention in the temporal rhythm" of matter. Descriptions of collisions in the accelerator repeatedly reference its capacity to simulate the conditions of the big bang. Films and literature at the lab speak of the Tevatron's ability to re-create the earliest conditions of the universe. Although this is not the literal remaking of the moment of creation, it is a simulation of that process. The accelerator and its operator-builder, with "various techniques, gradually takes the place of Time: his labours replace the work of Time."[14]

Fermilab describes itself in grand terms as "a remarkable creation of the human mind and spirit . . . dedicated to revealing the mysterious beauty and evolution of the physical universe."[15] The particle collisions made possible by the accelerator are woven into the larger story of a quest for a grand unified theory of physics, a strikingly similar motivation to the alchemist's pursuit of "an all-encompassing understanding of matter, energy, space, and time."[16] The linkage of space, individual experiments, and the unification of physical knowledge implies that something more than ordinary science is taking place here. The texts reinforce this interpretation by claiming that the work of the accelerator and the scientists who operate it eventually will "answer the deepest questions of matter and energy, space and time."[17] This sort of rhetoric began to circulate five centuries ago around alchemy.

The accelerator itself is the key focus of the laboratory's story. Rhetoric always returns to the possibilities that emerge from interaction of the machine and its users. Nowhere is this more apparent than in the lab's description of itself as a "scientific crucible."[18] The image of the crucible, taken from alchemy, can be interpreted in two ways. In one sense it refers to the use of the accelerator as an instrument that facilitates the manipulation of matter. In another sense, however, it signifies severe tests that determine one's admission to the community of scientists gathered at FNAL. The path to becoming an experimental physicist is itself described as an act of sublimation as an individual's character and talent are tested and reformed according to the requirements of the community. Thus certain experiences are reserved for those individuals who demonstrate a talent and reverence for basic science, a respect for the extraordinary nature of their work, and a willingness to submit themselves to the arduous training necessary to achieve success in the field. This process echoes both Eliade's and Spencer Weart's descriptions of the nature and significance of alchemy in modern science. Eliade's description, though not focused explicitly on rhetoric, draws our attention nonetheless to the creation of power within a community centered on technological experience: "This experience is their monopoly and its secret is transmitted through the initiatory rites of their trades."[19]

The Mundane and Sublime Experience of Time

High-energy physics at FNAL is depicted as taking place within a special liminal space and time, a space removed from the potentially destructive influence of the ordinary world. Thus the experience of visitors and physicists alike is defined in part by a distinction between ordinary and special time. The boundary between the world of the scientist and the nonscientist is not merely a matter of physical space or even identity; it is a function of the character of time-space each inhabits. Consider the following quotation from John Huth, a FNAL physicist: "I think support of this kind of institution is a measure of the level of a culture. People built huge cathedrals in the Middle Ages, when there wasn't much surplus money to throw around. If you take a certain amount of pride in your culture you build pyramids or cathedrals or monuments. Of course, what is fundamentally different here is that, rather than just building some huge monument, you are actually learning something in the process."[20]

Leaving aside both the startling bravado and historical inaccuracies of the statement, it remains a pithy example of recurring comparisons between the laboratory's work in high-energy physics and the great monuments of culture. Comparing the laboratory to a monument, we sense that time and space have been fixed somehow for the purpose of marking a particularly important set of people and their activities. Such articulations of special or liminal place and time pervade our experiences of both the natural and technological sublime. In the case of the natural sublime, our altered sense of time is linked often to some divine or sacred presence. In the case of the technological sublime, our sense of difference is focused on the technology itself. Simply put, both forms of the sublime signal the suspension of ordinary ways of being in the world.

The mere presence of the accelerator is awesome to many, and registers for some our deepest fears about the power unleashed through nuclear technologies. Yet the sheer experience of its presence is insufficient to produce a sublime response for visitors consistently. We must also comprehend its capability—the other worlds it might reveal to us—for the experience to take hold fully. Thus other supportive texts must be invoked to aid and guide our interpretation of the experience. In this sense the accelerator is indicative of the shifting cultural bases for the technological sublime. It is doubtful that the laboratory's earliest visitors would have missed the point, steeped as they were in the larger discourse of the cold war and the cultural stature of physics and poised as these were between the sublime and the terrible. Today, however, the technological sublime is as likely expressed in megabytes as in megatons and so the buildings, sculpture, grounds, and technology must work together to focus audiences' range of possible interpretations. Accordingly the grounds are designed to emphasize that the laboratory is set apart from the rest of the world in both purpose and structure. It is a self-contained world created for the express purpose of fostering scientific research into the origins of matter. The sublime nature of that task and the technology required to perform it suggest connection to a unique and more powerful reality than that experienced by those outside the laboratory gates. Although the technological sublime is not dependent on an association with the sacred, it is nonetheless driven by the distinction between ordinary and extraordinary. Even the dark specter of nuclear annihilation can be mustered for the purposes of making such a distinction. The "nuclear" has been and continues to be presented to the public as far removed from "the normal routines of everyday life, as something extraordinary and potentially dangerous, though its dangers can be domesticated and its potency harnessed to the needs of everyday life."[21] Understood from this perspective, the laboratory grounds are filled with objects that mark its special character—the prairies, the buffalo, the sculptures, Wilson Hall, and, most important, the accelerator and its related technology. These objects and entities are blatantly marked off and "stick out" both from the surrounding space of suburbia and the norms of scientific research facilities. As powerful reminders of difference from the ordinary, these elements demarcate the work of FNAL from other human pursuits.

This otherworldly sensibility has been a feature of the site since the laboratory's opening and has often served to facilitate relationships with the surrounding neighborhoods. For many of its suburban neighbors, the lab is an oasis of green and open land in a sea of tract homes. Differentiated from both the farmland that preceded it and the suburbs that surround it, it is presented as a primordial space: America before America. Thus the experiences of space and time at FNAL point back further still to fundamental states of matter at the beginning of time itself. Accordingly it is an odd theme. The lab Web site and other texts interpret the grounds through repeated reference to time and material states that have long since passed out of existence. Laboratory brochures extend this interpretation to the accelerator and its detectors, analogizing it to a "time machine" that re-creates the most fundamental states of the universe and simulates the conditions immediately following the big bang. The lab grounds have been reconstructed so as to reflect a similarly fundamental state of existence, a time of prairies and oak savannahs well removed from the agriculture and suburban sprawl that followed. The audience is thus transported in time.

Frequent comparisons between the work of the laboratory and the great cathedrals of Europe also underscore the laboratory's purposeful attempts to set itself apart from mundane experience. Within this frame the laboratory and the accelerator constitute more than public projects; they are monuments to culture and therefore justifiable without thought to expense, practicality, or sacrifice. The availability of money or the appropriateness of funding such a facility is seen as a mundane concern, perhaps necessary but not decisive to either motive or effect. Focus on the laboratory as a cultural achievement allows Fermilab to argue for its continued importance within the larger society. Indeed the symbolic construction of the lab's extraordinary nature and purpose ultimately serves important political, arguably very mundane, purposes. By both physically and symbolically removing itself from the realm of everyday existence, the particle physics community attempts to maintain a particular vision—its ideal relationship between science and society. This vision of science-as-a-monument-to-culture has been systematically deflated as a result of increasing government oversight, stagnant budgets, and shifting public interests. Yet stimulated by its own rhetoric, visual and verbal, the influence of outsiders—the government, critics of science, and environmental and religious activists—has served to solidify the laboratory's separationist rhetoric to its own denizens.

We have seen that FNAL is constituted as separate and special through the repeated articulation of the technological sublime across the lab grounds. Yet, at the same time, the lab's visual rhetoric refers to the experience of the natural sublime, which is expressed through the contrast between the stark primitive and the natural look of the lab's landscape in contrast to the congestion of the Chicago suburbs. Still, the unique state of the land and its distinction from that which surrounds it ultimately refers back to the science and technology of the laboratory. While the prairies and savannahs are extraordinary in relationship to suburban sprawl, the space revealed within the accelerator is fundamental and distinct in comparison to the world of ordinary physical experience.

In the midst of a highly developed suburban environment, the lab stands out as open space that is not only out of place, in that it is not developed, but also as a place "out of time" in its primitive manifestation as prairie and oak savannah. At the time of the lab's founding, this land was farmland, virtually indistinguishable from the surrounding area. Wilson's conscious decision to restore the land to its presettlement state reveals a key aspect of the laboratory's rhetoric. The grounds represent some primordial origin point coextensive with the research goals of particle physics. They mirror the work of the laboratory and its people, experimentally re-creating the earliest states of the universe in order to discover the underlying fundamental force that governs the functioning of all matter. Thus the distance between the mundane and extraordinary is represented for visitors through the repeated references to this transformation of "the prairie" and "the particle." Just as the grounds have been returned to their state of origin so that we can better understand the interaction of various natural systems, visitors are shown how the particles that constitute the substance of that world must be taken back to their most basic state to facilitate our understanding of how and why matter functions as it does.[22] While the laboratory's physical environment has evolved over time, to this day much of its rhetoric is still carried by its oldest structures.

Laboratory Directors, Ethos, and the Rhetoric of the Technological Sublime

Let us now turn to the rhetorical figuration of FNAL's directors as magicians of its sublime rhetoric. The rhetoric of the lab's built environment, public spaces, and texts was strongly shaped by the character and rhetorical performances of its first two directors, Robert Wilson and Leon Lederman. Each had the institutional authority to mark the lab according to his vision and in so doing merged his personal character with the evolving public image of the laboratory. While both Lederman and Wilson exerted undeniable influence on the laboratory, the rhetorical construction of their characters (ethos) in relationship to the lab is itself part of their influence. The technological sublime functions in part by linking technology to its creator. Indeed the strong association between sublime experience and the presence of a creator-builder concentrates political power within cultural institutions, organizations, and individuals. Long after their terms of service were concluded, Wilson and Lederman continue to function as key characters in Fermilab's narrative. Their figuration as builders is in fact key to understanding the technological sublime at Fermilab.

To hear them described by current and retired laboratory employees, one gathers that Wilson's concern was overwhelmingly with the construction of sublime spaces and technologies whereas Lederman's concern was to reassert and reinforce the identity and authority of the experimenter within scientific culture and the society at large. Richard Orr compared the two directors in an issue of *Ferminews*. Of Robert Wilson, he commented: "I think he was an artist almost before he became a physicist. Telling him how to run a laboratory was like telling Picasso how to paint. He stamped Bob Wilson all over the 6800 acres of Fermilab. This place looks like it does, feels like it does, is like it is, and the staff is like it is, because of Bob Wilson. His heart and soul are in every square inch of this place."[23]

In contrast, Orr describes Lederman in the following terms: "Leon put a world-class cachet on our experimental program. Educator, lecturer, author, leader—Leon was also a great diplomat, and he had to mend a lot of fences after Bob. . . . Having him as our director increased our standing quite a bit. We were almost a great laboratory waiting to happen, but we had to have a strong experimental program."[24] While the changing culture of the contemporary laboratory and its evolving relationship with the DOE have gradually eroded the power and influence of laboratory directors, Wilson and Lederman are still heralded as the men who founded the laboratory. The following sections consider the role of laboratory directors as both authors and performers of the rhetoric of Fermilab.

The Wilson Era: Art, Architecture, and Grounds

If we accept Orr's description, Robert Wilson was largely responsible for the sublime infrastructure that still marks the laboratory today. In fact, although Wilson exerted clear influence on both the process and the design, many people contributed to the construction of the laboratory. Yet organizational and historical narratives persistently and overwhelmingly portray Wilson as the chief architect and designer of the site.

Wilson's background is significant, particularly as it lends itself to reinterpretation through the technological sublime. Wilson came to the laboratory with a background in accelerator design and building. He started his physics career at Lawrence Berkeley National Laboratory, an environment that emphasized experimentation in both research and technology. While at the Berkeley Rad Lab, Wilson worked as one of many young physicists trying to iron out the fundamental challenges of accelerator design. His tenure there was marked by significant developments in accelerators and in basic physics under the direction of the legendary Ernest Lawrence. Wilson's own work was in the area of particle beam focusing until it was interrupted by the war. He joined the team at Los Alamos and worked on one of several teams charged with preliminary work for elements of the atomic bomb. Wilson's work at Los Alamos is often portrayed as tangential to his developing research program. Yet his experience there proved significant to his later work as director of Fermilab. As his many writings detail, he was convinced early on that basic particle-physics research needed to be distanced from the goals of the military-industrial complex. He resisted the secrecy and restrictions placed on the scientists at Los Alamos and at the end of the war gave up his security clearance and returned to Cornell to resume his academic career.

Wilson's experience at Los Alamos also gave him his first exposure to the relationship between the fortunes of organized particle-physics research and the power of sublime experience. Like many of his compatriots, Wilson witnessed the public's simultaneous attraction to and repulsion from the bomb and the power it represented. His experiences during and after the war were central to the rhetoric that emerged eventually at Fermilab. Wilson had a deep need to rehabilitate physics and reconstruct the laboratory as a space separate and apart from the military. The majority of his rhetoric was built in direct opposition to commonly held perceptions about publicly funded physics. Grounded in the sublime experience of the nuclear, he attempted to renegotiate the terms of physics' Faustian bargain with the federal government. The laboratory he envisioned was *not* Los Alamos. It was not secret, guarded, fenced, or inaccessible. It was *not* devoted to destructive applications of physics. It was *not* a weapons lab.

During his time at Cornell, Wilson earned a reputation for designing functional accelerators that were both efficient and relatively inexpensive to build. As he developed his skills as an accelerator builder, he also perfected his skills as a sculptor. A passion cultivated since his days at the Berkeley Rad Lab, this artistic hobby would prove to have a substantial impact on his work in accelerator physics and the then-emergent rhetoric of high-energy physics. His rhetoric resonated with the admonition that machines could be both elegant and powerful and that overdesign was both an aesthetic affront and a waste of time, money, and energy. He championed simple accelerators with straightforward operating systems that could be pushed to maximum energies.

Wilson's perspective on design had an early impact on the development of the NAL, particularly when he challenged the proposed Berkeley design for the NAL accelerator. He complained that, as proposed, the accelerator was overdesigned, and he argued that it could be built for less money and completed in less time if substantial design modifications were made. The decision-making panel was initially unimpressed, even annoyed, by Wilson's challenge to the approved design. After Edward Lofgren refused to build the Berkeley design at the chosen Illinois site, however, the AEC turned reluctantly to Wilson. Before accepting the post, Wilson demanded complete autonomy and authority to direct the facility and execute the design as he saw fit. After having received the necessary assurances from the AEC, Wilson made good use of his opportunity. Not only did he test his approach to accelerator design on a large scale, but he also set out to build a laboratory that was very distinct from other facilities in the national system. Wilson was determined to demonstrate that an accelerator could be built efficiently and creatively from the ground up, an attitude that challenged many of the established boundaries that then circumscribed accelerator physics. Thus he positioned himself as the builder-creator, a performance that would not only serve him in the short-term funding battles to come, but would become central to Fermilab's own creation story.

Wilson's rhetoric also encompassed the social environment that would be created at the laboratory. He spoke often of a scientific utopia, a protected space for the practice of science. His lab would demonstrate for employees and visitors alike the natural integration of physics and humanism that guided him in his work and that he believed could reconstruct the existing boundaries between the national laboratories and the public. Wilson's goal was nothing short of a complete revisioning of the laboratory as a professional, social, and cultural institution. Although the Stanford Linear Accelerator (SLAC), the NAL predecessor, had also sought to distinguish itself from the military-industrial complex by cultivating a complex but nonetheless strong relationship with Stanford University, Wilson's vision was even more radical. FNAL would stand alone at the frontier of physics as an independent laboratory.[25]

To accomplish his goals, Wilson paid particular attention to the physical environment of the lab as a means to fulfill his own sublime vision of particle physics. While conscious of monetary constraints placed on him by the AEC, he argued that the laboratory should be a visually dramatic environment that would demarcate the facility from the surrounding agricultural areas and create an atmosphere in which the scientists and engineers working on the accelerator could be at their most creative. Thus Wilson's rhetoric drew heavily on associations between nature and technological power that are the foundation of the American technological sublime. Some of the lab's most striking features and most powerful texts were born out of a carefully and frequently articulated philosophy that art and science were fundamentally unified. He argued that the beauty and power of scientific discovery should be reflected in the facilities and instruments that give rise to such work:

> I have always felt that science, technology, and art are importantly connected, indeed, science and technology seem to many scholars to have grown out of art. In any case, in designing an accelerator I proceed

very much as I do in making a sculpture. I felt that just as a theory is beautiful, so too, is a scientific instrument—or that it should be. The lines should be graceful, the volumes balanced. I hoped that the chain of accelerators, the experiments, too, and the utilities would all be strongly but simply expressed as objects of intrinsic beauty. Aesthetics is partly a matter of communication, and with so many people involved, I felt that everyone would appreciate the economy of good design and would keep their designs equally clean and understood.[26]

This statement, taken from one of several personal accounts of Wilson's early work at the lab, indicates that the fundamental work of the particle accelerator was never far from his mind. His chief goal was always to develop a machine capable of extreme energies, a machine that would eventually smash particles together in collisions at near light speed. The technological sublime inherent in this activity provided a means to present the accelerator as beautiful and powerful at the same time. When pressed by Congress and decision makers, Wilson advocated power over conservative design. Throughout his tenure as director, he pushed the accelerator to new energies, eventually achieving more than twice the beam energy originally proposed. Wilson's interest in power served his rhetoric as well. He was well aware of the political and professional capital to be gained by promising and delivering more accelerator for less money.

Much of Wilson's rhetoric is marked by similar acts counterbalancing political and professional interests. Nowhere is this more clearly expressed than in the laboratory's built environment. Wilson made a conscious effort to create an environment that was visually distinct from its predecessors and competitors in the national laboratory system. To begin, the laboratory was designed to be completely open. Wilson insisted that there be no fences, no guard gates, and no armed security personnel on the site.[27] Responding in part to increasingly vociferous critiques of the relationship between organized physics and the military-industrial complex, Wilson felt that the new NAL should embody the distinction between basic science and defense-related research. He took great pains to place the new laboratory in direct and public opposition to the secretive and isolated institutions that characterized the majority of the system. In 1986 a reporter from *Science Digest* remarked on this distinction: "No less extraordinary than the physics at Fermilab is the architecture of the place, which distinguishes it from other government funded laboratories. No chain link fence surrounds the 6,800 acre complex, built on a cornfield 40 miles west of Chicago. The access road leads not through a gate but through two steel legs of a towering, three-legged sculpture called *Broken Symmetry*. . . . There's no guard at Fermilab to discourage you from freely exploring the grounds."[28]

Fig. 2. "Broken Symmetry"

Note that the author contrasts FNAL with two enduring symbols of the weapons labs that served the military—the chain-link fence and the guard gate. While it is impossible to speculate as to the origin of such an observation, the lack of fences and gates is referenced across numerous Fermilab texts, most notably Wilson's own writing. In place of the traditional guard gate, Wilson placed a sculpture fashioned from a scrap ship hull. "Broken Symmetry" (fig. 2) expresses the disjuncture between the observable disorder in the universe and the hidden order he sought to reveal through physics research. The sculpture appears disordered or unbalanced from all angles but one—directly underneath. Wilson hoped that as guests and employees pass through Broken Symmetry, they would glimpse the hidden order of the sculpture just as physicists have glimpsed the order of the universe through their work at the laboratory. Perhaps to contrast the turmoil that seemed to characterize the era of the lab's founding, Wilson made many such tangible displays of openness in order to initiate contact with the surrounding communities. Tours of the facility were begun during his tenure as director, and the public was encouraged to visit the site to view, among its other attractions, the herd of buffalo that had been relocated there.²⁹ As both an artist and a physicist, Wilson continued to perform his vision for the laboratory in both functional and decorative installations, combining the two whenever possible. To this end he constructed extremely large and complex sculptures, many from materials used in the construction of the accelerator and other facilities. The Obelisk (fig. 3), the Möbius Strip, and "Tractricious" (fig. 4) are among the most visible and obvious expressions of his vision. "Tractricious" provides a particularly interesting example of how Wilson's work is read through the lens of the technological sublime: "'Tractricious,' designed by Wilson and constructed by members of the Technical Support Section, sits in front of the Industrial Complex. The structure is comprised of 16 stainless steel outer tubes, made from scrap cryostat tubes from Tevatron dipole magnets, and 16 inner pipes from old well casings. Each tube is free standing, designed to withstand winds up to 80 mph."³⁰ This description of the sculpture reveals an emphasis not so much on the *form* of the piece, but rather its construction and inherent strength—it is simultaneously simple and complex, delicate and strong. When examining the construction of the piece and the physical strength of its components, we are drawn to the sculpture's relationship to the constituent elements of the accelerator and the ingenuity of the individuals involved in its design and construction.

Characteristic of the modernist discourses of the day, Wilson took great care to marry form to function in his designs for power poles, capacitor trees, exhaust outlets, and other ancillary systems necessary for the operation of the accelerator. With each new design, he was careful to reference the central activity of the laboratory and the technology through which it is performed. Here he describes how his habit of melding the sculptural and the functional began early in his career: "When I was a graduate student at Berkeley, I used to go into the lab at night when no one was there and construct *big, kind of scary* figures from whatever was lying around and leave them there for people to find the next day."³¹

Wilson's description reveals that, while he often spoke in terms of the beautiful, his visions had their roots in the sublime. His preoccupation with "big, kind of scary figures" reveals the link that *Science Digest* describes: an "eccentric aesthetic—at once sculptural and mathematical—dominates the architecture of Fermilab."³² Wilson continued to interpret his work and its meaning for the laboratory throughout the course of his life. He authored several articles on the topics of founding and designing Fermilab and facilitated the scholarship of others in this area. These texts work in concert to produce a remarkably consistent historical narrative that emphasizes Wilson's own role in the creation of the laboratory and consequently obscures

Fig. 3. "The Obelisk"

the influence of outside actors and rhetorics. There are few references to work done by other engineers or architects.

Furthermore Wilson was clearly influenced by the high modernist sensibilities of the time. He was in that sense constructing a laboratory that articulated with the dominant architectural ideologies of the times. While many of Wilson's designs are self-referential, they also incorporate the vocabulary of the International Style. Nonetheless laboratory narratives depict an independent visionary who assumed that others would be both attracted to and bound by his vision for the laboratory. Here Wilson characterizes his relationship with the architects working on the project: "The architects didn't quite know what to make of me. I was their client. . . . but I would not hesitate, as a sculptor, to criticize their aesthetic forms, nor would I, as a

Fig. 4. "Tractricious"

physicist refrain from calculating strengths of beams and flows in pipes.... Tearing up plans and then jumping up and down on them seemed to get my point across."[33]

Wilson's obsession with the architecture and his autocratic management style is evident in his larger vision of the laboratory grounds. He insisted on an environment devoid of the temporary-looking, boxlike structures of "institutional gray" that populated other laboratories, particularly Los Alamos. To facilitate a dramatic break with the past, he utilized interesting and unusual forms and also dramatic, disruptive color patterns. At the suggestion of Angela Gonzalez, the lab employed patterned combinations of orange, blue, and yellow to designate certain areas and facilities. To this day the detector facilities and some village houses are painted in various combinations of orange and blue, and the helium tanks that supply the accelerator's cryogenic systems are arranged in a gradient from light yellow to orange.

Wilson Hall, the main high-rise office building on the site, is perhaps the most dramatic expression of Wilson's attempt to construct the laboratory as a dramatic space. A thoroughgoing modernist structure, Wilson Hall closely resembles the Ford Foundation building in New York. The poured concrete building soars fifteen stories and features a central atrium area banked by poured concrete "buttresses" on either side.[34] The central interior space is open to roof skylights and windows extend from floor to ceiling on the front and back of the building. The profusion of light produced by this design blurs the visual boundary between the technological and natural. The entrance area features large planters with changing seasonal displays and a large brass Foucault pendulum suspended from the top floor. A general dining area and cafeteria occupy the remainder of the first floor atrium area. Crossover balconies on the second and third floors house the laboratory's art gallery and library, respectively. Offices are located in the supporting side structures, arrayed along long hallways parallel to the sides of the building.

As a laboratory building, Wilson Hall is a unique structure. It soars above the prairie landscape. Its scale and inherent drama seem to reference the importance of the laboratory's work. Wilson's intentions with respect to the design are clear and well documented. He invited submissions from a number of architects and encouraged them to create designs that defied the architectural conventions of the government. The chosen design, while actually more conservative than other proposals, was without rivals in the national laboratory system, given its height, curved "buttresses," and seeming lack of supporting structures. Even the Foucault pendulum located in the center of the atrium defies explanation as it seems to descend out of thin air with no visible anchor. Unlike other central laboratory buildings, the structure was not designed to be hidden from view but instead rises dramatically from the wooded horizon when viewed from a distance. A number of Angela Gonzalez's drawings (figs. 5–7) reinforce this visual interpretation of Wilson Hall. The artwork expresses the union of nature and technology through the merging of images of the accelerator and Wilson Hall as manifestations of the sublime.

The artistic, architectural, and technological elements built and installed during this period served Wilson's larger vision of a scientific utopia that would be realized in part through confrontation with the sublime. His comments indicate the degree to which he understood and capitalized on the

FIG. 5. "Wilson Hall—the Power of the Sun." Drawing by Angela Gonzalez, cover for the booklet *Starting Fermilab*

persuasive appeal of aesthetic experience for both scientific and nonscientific audiences:

> These early meditations of mine were often a kind of fantasy in which I envisaged the Laboratory as a utopian place where physicists coming from all parts of the country—and from all countries—would be doing their creative thing in the ambiance of well-functioning and yet beautiful instruments, structures, and surroundings that would reflect the aesthetic magnificence of their discoveries and theories. My

FIG. 6. "Wilson Hall—Moonlight over the Prairie." Drawing by Angela Gonzalez, *Starting Fermilab,* p. 16

fantasy of a utopian laboratory clearly required a setting of environmental beauty, of architectural grandeur, of cultural splendor. . . . My justification [for the expense involved] was that if we produced a dowdy site with shabby buildings, then the technical people we wanted with us would not come and the statesmen, who might judge us in part by appearances, would not, in the long run, give us the funds we would need for our physics.

Time and changing discourses have taken a toll on Wilson's creation. New accelerator technologies have gradually changed some of the landscape. The interior space of Wilson Hall has been modified to include a new reception

Fig. 7. "Wilson Hall and Obelisk Reflected in Pond." Drawing by Angela Gonzalez, *Starting Fermilab,* p. 25

desk, exhibit spaces, and a new wood-paneled conference room.[35] The terrorist attacks of 11 September 2001 and the encroaching suburban neighborhoods have had the most profound impact on the laboratory. The main entrances to the laboratory are now blocked by control gates to discourage both terrorists and commuters in search of a way out of rush-hour bottlenecks. Security has once again become a dominant feature of laboratory life as Fermilab contemplates its own vulnerability. The shift has been profound for both those who work there and those who seek to visit. Visitors to the facility must now stop and sign in if they wish to leave their vehicle to visit Wilson Hall. Once there they are restricted to the first floor unless escorted by a laboratory employee. Despite these many changes, however, the fundamental elements put in place by Wilson—the prairies, buffalo, sculpture, and basic facilities—remain to be seen and reflect the rhetorics that shaped the early years of the facility and the interpretive boundaries that mark the space of the laboratory.

The Lederman Era: Establishment, Education, and Ethos

In 1978 Robert Wilson resigned as director of the laboratory, discouraged by the government's initial refusal to fund the accelerator upgrade, the energy doubler. As successful as Wilson's low-budget, high-drama approach had been at convincing influential government officials that the facility could and should be built, his management style could not sustain the laboratory when funding crises began to surface once again. Despite impressive achievements such as regular upgrades in energy and significant discoveries, notably the discovery of the bottom quark, the federal government was backing away from its promise to fund an upgrade to a 600–1,000 BeV machine. Those outside the lab pointed to a growing recession driven by skyrocketing inflation and interest rates as an obvious cause for the government's hesitancy. However, those within the physics community still felt strongly that they could and would convince the government of the necessity of the project and do so in much the same way they once influenced the creation of a postwar research program. To this end the lab turned to one of the architects of the postwar science policy, Leon Lederman.

Whereas Wilson was cast as an accelerator builder, Lederman would play the role of politician and relationship builder. As an influential member of the physics community, Lederman represented the physics establishment and embodied the "best science" model revered by Vannevar Bush and his followers. He was young, smart, energetic, and unafraid to engage with the politics of big science. The most striking difference between Lederman and Wilson was that, while Wilson had always positioned himself as an outsider and a bit of a renegade, Lederman had learned early the power of being an insider and worked to establish positions of influence within the government and the field of high-energy physics. This shift mirrors the lab's rise to prominence within the field of high-energy physics. No longer an outsider facility, it required an insider director. Lederman had been a member of the committee that initially reviewed and responded to the Ramsey report that recommended the 600–1,000 BeV accelerator and thus was uniquely prepared to reiterate and reinforce the arguments that had been offered in support of the machine. His presence also reminded those in power of his role in originating the idea of the "truly national laboratory," the inclusive model of laboratory structure and management upon which FNAL organization was based.

As a scientist Lederman had credentials that were ripe for cultivation in the laboratory's rhetoric. He had been a member of the team responsible for the discovery of the bottom quark and as a result of this work had earned the National Medal for Science and the Nobel Prize for physics. The prestige associated with these accomplishments reinforced the community's belief that Lederman epitomized the balance of intelligence, accomplishment, and political savvy they sought in their chief representative. Though some of those at the lab remained loyal to Wilson and protective of his legacy, it was widely believed that Lederman, by virtue of his position among the political and the physics elite, was better equipped to lead the now-established laboratory into the next decade and toward new energy frontiers.

Lederman used his prestige to great rhetorical effect. He mobilized his power network to lobby influential federal officials in favor of upgrades at FNAL and the general expansion of particle physics. He succeeded at the initial task set before him and secured funding for the energy doubler shortly after he assumed the directorship. The doubler would be pushed to greater and greater energies, reaching 800 BeV in 1984, 900 GeV in 1986, and approaching 1 TeV in the late eighties. Having reached the unprecedented "TeV" mark, the doubler thus came to be known as the "Tevatron." Although the upgrade of the accelerator was certainly the crowning achievement of the era, there were other significant technological improvements that also figured heavily in the lab's development and success under Lederman. Development of the antiproton source began in 1983 and work on the lab's collider detector facility (CDF) was validated in 1985 with the first observation of proton-antiproton collisions at 1.6 TeV center-of-mass energy. Work on a second, complementary detector facility, D0, also began during this period. In short, although Wilson oversaw the immense task of building the lab and its technological foundation, Lederman supervised an expansion of lab infrastructure that would prove critical to its establishment as the world's leading center for particle-physics research.

Lederman's underlying goal was to reassert the dominance of particle physics in the U.S. program of federally funded research and development. Although Wilson's tenure was marked by efforts to unify disparate groups in support of high-energy physics, Lederman's task was to continue to distinguish the field from its competitors and to attempt to reestablish the independent authority of high-energy physicists in the power structure of federally funded research. Wilson made concerted efforts to gather personnel, to design and build the accelerator and necessary support facilities, and generally to begin to forge (quite literally) a unique sense of identity for the laboratory. In contrast the eighties were the time during which the lab assumed and solidified its leadership role in the field. Now that the facility had broken new ground in terms of geography, user access, and management structure, the next task was to shift attention to the energy frontier and U.S. dominance of particle physics in the world of basic research. Within five years of assuming the director's post, while still supervising the upgrade of the existing facilities, Lederman was lobbying for the construction of the superconducting super collider (SSC) in Illinois.

One of the chief complications of Lederman's tenure was the rapid maturation of the programs and facilities at CERN (European Organization for Nuclear Research), the European research facility in Geneva, Switzerland. Between upgrades in power at FNAL, CERN's accelerator briefly laid claim to the title of "World's Most Powerful Particle Accelerator," a development that signaled the viability of European competition. A number of significant discoveries by CERN physicists reinforced this perception. Physicists at FNAL began to think of the race for the energy frontier in global terms and thus worried not only about distinguishing themselves from other branches of physics, but also from competing facilities in other countries.

As a result of these changes, the technological sublime came to be abstracted from the site and facilities at FNAL. The focus whipsawed back and forth between existing technology at FNAL and technology yet to be developed for the SSC. Over time the sublime was redefined in terms of as-yet-unachieved energies and articulated through a discourse of upgrades and "next machines." The fate of initiatives to build the superconducting super collider provides informative examples of this process. The SSC was not promoted solely in site-specific terms, but rather as another national facility. The uncertainties associated with the site search left Lederman and others to mount two sometimes conflicting rhetorics. On one hand those at FNAL felt strongly that the new facility should be located in Illinois and make use of the FNAL's existing facilities. But linking the SSC too strongly to any one site might mean risking the whole project, and so advocates often recast the drive for the new accelerator in global terms that disassociated the machine from any one contender site. Instead of positing the SSC as yet one more accelerator in the cascade of machines already in place at FNAL, the U.S. physics community focused its attention on the "imaginary" sublime of the next machine, wherever it might be built. This shift meant that the rhetoric at FNAL began to revolve around the machine (and the site) to come rather than the one already in existence.

In anticipation of this progression from an existing to imagined machine Lederman launched a very public campaign to promote scientific literacy, "scientize" the public, and create a constituency that could imagine the energy frontier. Whereas Wilson had sought to create a sublime object and environment that would resonate with the diverse audiences he needed to unite, Lederman sought to create audiences suitable for the sublime object he and his community wished to create—audiences that would imagine, understand, and advocate for the SSC regardless of its location. Always a staunch advocate of growth and reform in science education, Lederman revealed his larger political agenda, "physics first," in his science-education programs.[36]

While Lederman argued repeatedly that scientific literacy would reap the widespread benefit of creating an educated and rational citizenry, he simultaneously advocated the expansion of an elite class of decision makers trained in scientific practice. In this he was following the model of science-society relation that Vannevar Bush had initially set forth in 1945 in his report to the president, *Science: The Endless Frontier*. The education programs developed during this period eventually were classified according to a two-tier system that reflected the influence of Bush's "best science" model on leaders of Lederman's generation. Second-tier programs were geared toward educators and parents and were designed to enhance existing science-education programs. First-tier programs and initiatives involved scientists in the control and management of science education programs. These programs, including "Saturday Morning Physics" and the establishment of the Illinois Math and Science Academy, were distinguished from other initiatives in part by stiff entrance requirements for participants. Entrance requirements included not only top-level grades and scores on standardized exams, but also recommendations by teachers and administrators. This emphasis on starting with the "cream of the crop" hints at Lederman's latent desire to identify potential initiates for the physics elite. On close inspection the programs developed during this period were at least as much about replenishing the ranks of physicists and reinforcing scientific rationality's claim to cultural and intellectual dominance in the United States as they were about democratizing decision making in science policy.

At the same time, however, Lederman's emphasis on creating the next generation of physicists led to the emergence of a distinct "experimenter rhetoric" that shifted the discourse away from the builders valorized during Wilson's era and emphasized instead the users who populated the laboratory's many experiment collaborations. In order to establish the authority of particle physicists, particularly American particle physicists, Lederman set about defining the experimenter as distinct from other physicists in high-energy and also those pursuing other types of physics research. Despite the realities of a field in which far more people get paid to *do* physics than to *think about* physics, Lederman sought to liberate experimentalists from archetypal images of Einstein, blackboards full of equations, and serious-looking men deep in thought. The challenge for Lederman and those who would follow was to distinguish the experimenter from the theorist without sacrificing the status, prestige, and social power accorded to the theorist.

Lederman combined his personal experience and best-science mentality to make the lab a testing ground for scientists and scientific principles alike. But Lederman's rhetorical style was substantially different from his predecessor. Whereas Wilson would frame the builder in dramatic terms, Lederman articulated the experimenter through humorous images and anecdotes that soon became his trademark. A pithy example of this can be found in his book, *The God Particle*, where he makes a pointed attempt to distinguish between the theorist and the experimenter: "Experimenters don't come in late—they never went home. During an intense period of lab work, the outside world vanishes and the obsession is total. Sleep is when you can curl up on the accelerator floor for an hour. A theoretical physicist can spend his entire lifetime missing the intellectual challenge of experimental work, experiencing none of the thrills and dangers—the overhead crane with its ten-ton load, the flashing skull and crossbones and DANGER, RADIOACTIVITY signs. A theorist's only real hazard is stabbing himself with a pencil while attacking a bug that crawls out of his calculations."[37]

Notice how the distinction is defined specifically in terms of behavior, denying oneself sleep or exposing oneself to danger. The technological sublime subtended by danger, contra Wilson, is thus evident in description of the experimenters' relationship to the machine. This description, offered in the context of a larger text explaining the principles of particle physics and advocating for the SSC, is designed not only to distinguish the experimentalist from the theorist but also to associate the experimentalist with the sublime instruments of particle physics research. In this sense the experimenter becomes the new symbol and interpreter of the technological sublime. If Wilson's focus was the construction of a sublime technology, Lederman's goal was to reinforce the unique relationship between the machine and its users.

The experimenter-accelerator relationship remains a central element of the lab's rhetoric. Yet, in part because of the perilous flirtation with danger created by Lederman's version of that technological sublime, it has been challenged recently by the presence and work of health and safety personnel. One physicist interviewed during this study expressed frustration with what he regarded as challenges to the experimenter identity and authority. Here he distinguishes between work in the "cowboy" days and the manner in which health and safety issues are handled today:

> We had a six foot diameter pipe that was a hundred and fifty, two hundred meters long. I wanted to put a penny in the beams box, so that I could calibrate the software, so I brought the pipe up to here,

opened the window, climbed in, went down the pipe, put the penny in, crawled out. . . . I broke more safety regulations in doing that . . . but it was fine that we did it. [If] I did any of those things now, I would have the weight of everybody falling down upon me. . . . That violates enclosed space regulations—somebody would have had to survey the pipe [for] radiation, [do] radiation surveys. It's absolutely idiotic! As a physicist, knowing what that pipe has been through, the idea that I would survey that pipe for radiation hot spots is so stupid . . . I mean it's just crazy!

For most people, monitoring of exposure to radiation seems like a common-sense precaution. But, like the alchemist and crucible, the experimenter assumes he has a proprietary relationship with the technology he uses. To interpose safety regulations, expressions of a profane world where control must be managed through outside entities, was thus to violate the sacred nature of the bond between a researcher and his instruments.

Traditionally lab directors have managed outside interference through discourse that tended to protect the autonomy and authority of users within laboratory culture. Throughout his tenure as director, Lederman fought to keep the experimenter at the center of the laboratory's rhetoric. There was much work to be done. Fermilab's newly formed detector collaborations, CDF and D0, were growing rapidly and, if completed, the SSC would have been home to some of the larger researcher communities and collaborations in the history of physics. The careful cultivation of a user-centered culture seemed necessary for the success of both the existing laboratory and the proposed laboratory. In the wake of the SSC cancellation in 1993, however, these insular user-centered rhetorics grew to include stakeholders from outside the laboratory culture as high-energy physics confronted changes necessitated by shrinking federal research budgets. The move away from small fixed-target work to large collider-detector collaborations allowed the physics community to "do more with less," but the change also meant significant upheavals in the pedagogy, sociology, and rhetoric of high-energy physics as the lab's rhetorical boundaries became increasingly permeable to public critique. The following chapter documents the impact of flat budgets and the shift to large detector collaborations through the stories of those who work at FNAL under these latter-day conditions.

FOUR

Practice and Perception

As I study the laboratory and its meaning, I talk with those whose lives are, in a very real sense, defined by the machines that dominate life in this community. I've toured the detector facilities with experimentalists whose work is dependent on the optimal functioning of any number of technological systems. I've walked the grounds with accelerator physicists who build and rebuild the many accelerators that, acting in concert, enable the experimenters to do their work. And I have spent time with staff members and administrators who must work to convince outside audiences of the necessity of these machines and those yet to come. In a very real sense, all of these people are here because of the machines. Their professional life and world is built around the continued health of the accelerator and its detector collaborations.

Not surprisingly, laboratory employees often tell me stories of their life with the machines. They tell me what it means to work at such a facility. They talk to me about their attraction to science and technology and the world that each represents. They emphasize that physics without experiment would be little more than philosophy. As I listen, I wonder what happens when I become an audience for their stories? What do these tales tell me about them? How does the sharing of their stories give both meaning and structure to our relationship with each other and to our positioning within the space of the laboratory? How does the articulation of identity in relationship to the science, the technology, and the institution call into being the very boundaries that separate my world from that of my companions?

Identities in Transition

Boundary rhetorics is a term used by rhetoricians of science to mark a scientific community's attempt to stabilize practices and institutions from within and to resist the "corruption" that is assumed to come from outside influences. While accurate, this way of putting it frames the phenomenon of erecting boundaries in adversarial terms. To think of the world as carved up by fundamentally negative boundaries is an equally negative rhetorical strategy and often misses the point. Boundary rhetorics are as much about claiming one's own identity as rejecting another's. Seen from this perspective, boundary rhetorics are both constitutive as well as strategic. The sublime narratives of physics do indeed produce the strategic effect of dividing and disciplining the social world into categories—science versus nonscience, science versus politics, theory versus experiment, "big" science versus "little" science, science versus technology. But they are also the glue that holds the high-energy physics community together as a community. They confer identity. It is in fact the balance point between identification and division that produces the boundary itself. We cannot hope to understand the laboratory's communication with outside audiences unless we first understand how discourse and rhetoric shape life as it is inclined *inside the* laboratory.

With these reflections as our starting point, we might question the degree to which scientists are aware of their own rhetorics. It is fairly easy to assume that scientists are unaware of the influence of their own professional discourse. And yet there is much evidence to contradict such a claim. Anyone who spends time in any national laboratory will be struck by the degree to which many physicists and employees can account for the role of rhetoric in their enterprise. They theorize and strategize. They critique the rhetorical exploits of fellow scientists. They read social studies of science and react with amusement, derision, questions, and concerns.[1] Faced with regular demands to justify their enterprise and stake out their territory in reports, presentations, white papers, and the like, physicists working in the national laboratory system are often keenly aware of the impact and importance of their own rhetoric.

The rhetoric of scientific institutions therefore reveals both the centripetal and centrifugal forces that construct the rhetorical boundaries of the laboratory and give meaning to the divide between science and society. Boundary work is centripetal in the sense that it builds and binds the laboratory community through the development of common practices, experiences, and values. This discourse functions in a subtle but powerful manner to construct the world within the laboratory—the sense of community that pervades Fermilab; the technologies that define its research practices; and the institutions and the symbols intended to exemplify its values. At the same time the centrifugal forces of discourse pull the laboratory outward into the larger culture. Boundary rhetorics, even when they confer solidarity on insiders, must also manage the world outside by placing the laboratory in relationship to other social practices and cultural institutions—the standards of suburban life, for example, or debates about the allocation of taxpayer dollars to scientific research or the oversight of the government. The rhetorical boundaries of the laboratory emerge from the need to balance or stabilize competing social and discursive processes. The boundaries, I contend, are made visible through images, metaphors, and narratives that negotiate individual, professional, and institutional identities in the face of material and ideological limitations of history, technology, and culture.

Such is the case at FNAL. When the laboratory was founded, sublime rhetoric served symbolically to purify the practice and products of physics research by blending mystery and magic with the rationalist frameworks of modernism. This rhetoric not only met the identity needs of the physics community, but also responded to critiques of the military-industrial-academic complex in the late 1950s and 1960s. In later years the upheavals of the SSC cancellation and growing uncertainty about the future of the field signaled the need for a concerted effort to redefine high-energy physics and stabilize its position within the larger culture. In many ways that project of stabilization is still underway as the community looks to its history in an attempt to reinvent its future.

In this chapter I investigate the "inside-outside" rhetorics of Fermilab by way of interviews I conducted with members of the Fermilab community, graduate students, physicists, and staff members from various divisions at Fermilab and occasionally other laboratories. These conversations occurred during—and so mark—a significant period in FNAL's history: the end of high-energy fixed-target runs, the shutdown of the "old" main ring, and the transition to a new era of research inaugurated by the completion of the main injector. At the time most of these interviews were conducted, the last high-energy, fixed-target run had been completed, and the community was looking toward a long shutdown period necessary to upgrade the collider detectors and complete work on the main injector, the recycler, and the NUMI project.[2] Emotions in the community ran the gamut from excited anticipation to mournful recognition of the definitive end to high-energy fixed-target work. For some it was the beginning of a new and better future for the laboratory. For others it was the end of an era.

The end of high-energy, fixed-target runs held a special symbolic significance for many, even those not officially associated with fixed-target experiments. These experiments were generally done by much smaller collaborative groups than those that had grown up around the CDF and D0 detectors. While many of these experiments were staffed by several dozen physicists and graduate students, they were small in comparison to collider detector collaborations involving several hundred scientists.[3] The shift to large detector collaborations had been set in motion by a gradual flattening of funding for high-energy work. In short, with fewer dollars to spread around and a perceived need to explore particle interactions at increasingly high-energy regimes, physics moved toward bigger and bigger collaborations. The significance of this shift has been detailed in a number of sociological studies of the field, most notably Sharon Traweek's *Beamtimes and Lifetimes* and Peter

Louis Galison's *Image and Logic*.[4] The fact and nature of such change is not the issue here. What is instead the issue is the degree to which the community was actively processing this shift in its rhetoric. While such change can appear natural, inevitable, or merely psychological, it is in fact negotiated through discourse. The political economy of contemporary physics research demands that choices must be made among competing research priorities. The arguments leading to such choices and the justification that follow them can be articulated only through rhetoric. The identity needs of the community require that such change be integrated into the overarching professionalizing narrative of the field. The community needs to make sense of the counterintuitive and arguably counterproductive developments in high-energy physics by making it appear natural, normal, and continuous with the past.

Many physicists believed that the trend toward ever larger machines and research collaborations was potentially detrimental to the future of the field. Acknowledging the recent discoveries in the field and struggling to sort through the professional and political processing driving the change, some physicists worried aloud about the limits to growth and the impact of large collaborations on graduate education in physics. In an attempt to resolve their concerns, many of the physicists interviewed drew on their internalized narratives of the history of the field as expected. For some these changes were the result of identifiable and often ineluctable shifts in the political environment. For others change had resulted from strategic choices made by those within the community. Still others focused on the changing context of scientific research itself as problems in the field changed. Common to all of these responses, however, was the apparent *need* to explain the shift in historical terms, resulting in narratives that were, as one participant suggested, "doomsday stories" about the history and future of the field. While the identity of heroes and villains differed from narrative to narrative, almost all foretold the gradual demise of high-energy physics should the present set of power relationships prevail. To halt this downhill slide, most called for internal and external renewal of the field—a revival grounded in a return to the basic principles and practices that had first attracted them to work in high-energy physics.

"Doomsday Stories": Questioning the Future of High-Energy Physics

Working in a discipline with controversial origins, high-energy physicists have always exhibited certain anxieties about the future of their field. A great deal of this undoubtedly results from the struggles of the cold war, an era during which powerful physicists struggled simultaneously to distinguish physics research from weapons work and to maintain high levels of federal funding and support. From the perspective of many physicists, continued research support was the reward for their community's accomplishments during World War II. The government sometimes saw the situation differently, however. While government largesse continued to feed many laboratories, there were ongoing battles about the increasingly abstract nature of the work being done and the ever-increasing budgets required to sustain it. Thus the "future of the field" became a matter of almost constant negotiation, subject to reinterpretation year after year, project to project, and, most important, budget to budget.

While expressed in different ways by different participants, most participants located the origin of their field in World War II and the first large-scale, federally funded collaboration of the war. Depictions of the evolution of the field from that point to the present day differ widely, but a majority of those interviewed emphasized that high-energy physics may have reached the limits of its natural development given current financial and intellectual resources. One senior physicist at FNAL—who was a long-time scientific staff member and veteran of many accelerator upgrades—narrated the field's history this way:

> I have a doomsday story to tell you. . . . Back before WWII European science was the best science in the world. There were a couple of really good professors and they had their little group and they'd do stuff in their lab. World War II came along and the bomb came along and the concept of organizing the entire nation into one big activity came up. So they gathered many of these people—some professors, some not, some from Europe—and they figured out how to do this technical thing. The war ended and for some reason Congress poured a whole bunch of money on the table. [This allowed] some of these labs to evolve a little more and address the kinds of questions being asked by high energy physics then. You had a little accelerator division in the physics building and so studies were made in the basements of all these physics buildings. These little professors became like little princes—they had their funding from the government, they had their graduate students, they had their assistant professors, so they were like little princes with their little groups. When they started to ask questions that were larger, the concept of a national lab really started to evolve—not just for building bombs, but for doing other kinds of physics. . . . So, the country evolved by asking questions of that particular scale. There's a lot fewer places now, common places where the equipment is. The problem we're having at the moment is that the evolution of the way we do this science has slowed down so that the time scale is no longer the time scale of a graduate student like it was for me. It starts to get

measured in decades or so. . . . So the place that we've arrived at is that the next big question high energy physics needs to answer is going to be a *much* bigger question.[5]

This lengthy narrative depicts a field that is nearing the limits of existing frameworks for funding and support. The central metaphor of the story is that of scale—the scale of funding required, the scale of support available, the scale of the questions asked, the scale of the equipment, time, and expertise required to answer those questions. As the speaker brings his story to a close, he calls for a collection of effort not simply to gather support but also to rethink the very technologies and modes of research employed in the field: "In order to do it, I personally believe that we have to have another one of these collections of effort. I don't see it happening. If you ask me where I see the field going, I see it coming to a halt. . . . It's the syndrome of taking your technology and pushing it to the limit, there's no breakthrough 'cause it's simply spending more money and making the 'bigger bomb.' "[6]

For this veteran of high-energy physics, the "bigger bomb" is not a weapon; it stands for the next hadronic collider. Despite having invested the majority of his career in development of this type of technology, he openly expresses his concerns about the artificial narrowing of physics research through overreliance on this single technology, large hadron colliders, and the sociology and methodology with which it is associated, the large collaborative experiment. It may not be surprising that an accelerator physicist, whose vision of science is shaped by questions about the physics of the *machine*, would frame the future of physics in technological terms. However, similar concerns were echoed by a number of other participants, accelerator builders and experimenters alike. Many participants expressed deep-seated concerns about the narrowing effect of increasingly large and expensive research technologies and projects. Acknowledging a certain type of growth and progress in the field, they worried that resources have become concentrated in fewer and fewer projects. The dominance of collider experiments over alternative experimental methods such as tried and true fixed-target research or newer, more speculative modes of research raised questions for some about the current quality of training as well as the diffusion of control and oversight of the experiment. As one experimenter explained: "Department of Energy loves big projects. I mean they love these things. They like things that they can budget and account and put price tags on. . . . They want the LHC [Large Hadron Collider], they want D-zero upgrades . . . The physics that these big collaborations do is very interesting and you have to do it, but I wonder if we're losing a lot by not having smaller efforts where people learn the basics so that they can do the big [work]."[7]

The perception of intense competition for human and monetary resources drives the move to large experiments, but in turn it also produces tension within the community, as those who would vest the future of the field in large collider experiments confront advocates for smaller experiments or more diversified research. As this experimenter put it, "collidercentrism" dominates today's high-energy environment: "People who have been on collider experiments all their lives have this idea that the rest of the people are little ants.... [They think] 'if they would just stop running these stupid little fixed target experiments, we could get all the resources we need' when, of course, all their good people were trained on those fixed target [experiments]."[8] As this and other quotations illustrate, the move to progressively larger colliders is a clear point of contention within the field. Participants worried that this trajectory threatened the very fabric of lab culture as it changed long-established patterns of training that emphasized the intimate connection between experimenter and machine and thus threatened the holistic understanding of physics research. Furthermore the fear that mediocre scientists might "hide within a large collaboration" has sparked a discourse in which physicists now distinguish between the authors named on a journal article and those "who actually did the work."

Such talk reveals underlying tensions between competing models of research. The move to national laboratories and large collaborations was done in part in the name of "democratic research." As conceived, the "Truly National Laboratory" was a jibe at the perceived elitism and parochial attitudes of the coastal laboratories.[9] And, yet, the "democratizing" of research must be articulated in terms recognizable within the meritocracy that shapes the culture and practice of American physics. Many feared that the widening field and growing size of experimental collaboration would overshadow individual achievements in the field. It seems that the image of the renegade genius so revered in American science is not easily reconciled with an experiment involving three hundred to four hundred people. One participant argued that the move to large collider experiments might threaten the creativity that defined the field and the laboratory: "It stifles the ability of young people who often have crazy ideas, one of which may work out! If you're in this huge environment, you don't get the chance to go off ... and do something interesting. The risks are so big financially that the bureaucratic stuff kicks in and they won't let you do something that's maybe a little off the wall. That's bad for the field."[10]

Participants also worried aloud that the trend toward large research collaborations promoted a culture that rewarded mediocrity over talent and initiative. Steeped in the established "star system" of the physical sciences, many of those interviewed—from graduate students to senior scientists—saw few ways to realize individual recognition within a large collider detector experiment. In fact many argued that, beyond masking the otherwise unequal contributions of members, large experiments actively bred an environment in which "one can actually get away with essentially doing nothing."[11] As one staff scientist put it: "There's an enormous number of mediocre and incompetent people. It frightens me that physicists are supposed to be so smart, because you look around and there's a lot of people who shouldn't be doing it. They . . . fill up the author list and then they don't do anything but sit around and object at meetings and make noise and suck up money."[12]

Such criticisms were echoed not only in other interviews but in meetings and seminars as well. Public presentations in particular seemed to provide an opportunity for scientists to create "alternative" author lists in which the whole collaboration would be noted, but those "who really did the work" were singled out for special acknowledgement, often on a separate listing.

The definition of "doing the work" seems linked in part to one's responsibility for overseeing the machine during both its construction and operation. As one graduate student commented, comparing what she assumed to be the experience of a large experiment to that of a small one: "I haven't really been on a huge experiment . . . but from talking to my friends who are on large experiments [I understand that] what I've done already is more than they will ever do. I have more responsibilities since there are less of us. For instance, when they have shifts, there are six people on a shift and you have someone who you report to . . . you just sort of fill a seat."[13]

This theme was reiterated by scientists working on large and small experiments. Their concern seemed to stem in part from the ways in which these large experiments threatened the carefully constructed experimenter ethos at Fermilab. Many argued that a member of a large collaboration does not face the same physical and intellectual challenges as those faced by a member of a small collaboration. Consequently members of large collaborations were often described as being less intimately connected to both the machine they worked with and the data they produced. Furthermore members of large collaborations were seen as less inclined to take risks. Many of those interviewed described the experience of pulling an overnight shift, focusing intently on the loneliness and the physical and mental exhaustion such work entailed. When compared to the "redundant" human systems of large collaborations—marked as they are by highly specific skill sets and multiple workers per shift—large collaborations do not appear to flow naturally from the "cowboy culture" of the laboratory.[14]

With so much concern, how and why has the transition to large research projects been carried out? The significance of this question registers in the resignation and determinism evident in responses of many of those interviewed. Although various scientists and graduate students used the interview as an occasion to express their misgivings, most spoke as if the move to bigger colliders and collaborations was somehow inevitable. They located the necessary cause for change outside the actions of the community. It was caused by the political economy of "big science." Contrary to popular accounts of scientists as apolitical creatures, this perception of being a victim of larger forces shows that many of these participants were only too eager to talk about the relationship between politics and science. They did not accord much agency to themselves or their professional community in the process. The invocation of politics was instead an occasion to talk of their own helplessness in the face of what they characterized as a complex and often alien process. A number of those interviewed would point to others in the community who "knew the ways of Washington," but even those individuals often pleaded ignorance and innocence when charged with the crime of political savvy.

This pattern shows that this construction was a conscious choice of members of the high-energy physics. The need to maintain the boundary between science and politics was clear—and to speak about it was a particularly effective means of doing so. The boundary was thus established through activating discourse rather than by avoiding it or allowing others outside the field to dominate it. The conversations often focused on naming the "other"—the politicians, the DOE, members of the White House administration—and assigning degrees of agency to the various players in the drama. With few exceptions FNAL employees *denied* their own agency and power and preferred to cast the narrative in terms of sociological and political evolution—the system was changing around them, and they were adapting.[15] This structural view of the field's sociology has important rhetorical implications. The tale woven by those interviewed was not focused on strategic choice but was instead strikingly deterministic, both reflecting and reinforcing what seemed to be a widespread sense of malaise within the lab. The shared narrative depicts a world in which desire for larger colliders is transformed into a *need*, both scientific and political. But such colliders require a large scale, often with international collaboration and funding, and thus create a situation in which neither the physics community nor the government can respond adequately. From this perspective science becomes a large-scale public works program subject to the demands of a wide range of interest groups—scientific and nonscientific. For many participants the SSC provided vivid evidence of this shift, as failure to solicit sufficient funds for the project from external sources proved a major barrier to its completion.[16]

Despite the air of resignation, many participants persisted in their efforts to identify the causes that prompted the transition to large colliders. One physicist's explanation invoked both the notion of sociological change *and* strategic choice: "I think that the future of the laboratory is . . . dominated by the sociology of large hadron groups. It's not the DOE. We chose to go that way. People go to the DOE and say, 'this is the best physics in the world' . . . and get the money. The young people look around and go, 'if I apply for a job in neutrino physics, my career's over, but I can get a job on CDF.'"[17]

Yet such acknowledgements even of strategic agency were rare in this small, tight-knit professional community. Most physicists interviewed preferred to depersonalize the move to large colliders and spoke of change as the result of external events and interests rather than the motivated action of the physics community. Such a narrative has a hollow and somewhat troubling and contradictory ring, however, as it forces participants to acknowledge the field's utter dependence on the political environment. One physicist commented: "The field's been through many transitions and it's going through another transition to even bigger experiments. It needs people, but it doesn't have any money, right? And because it requires lots of money to build accelerators and experiments, it depends too much on public opinion, that which is reflected to congressmen. That's one thing I think is very negative for pure science projects because it depends too much on what's going on in politics."[18]

In the end most of those interviewed cast the problems associated with large accelerators as the result of a failed political process in which science and politics were clearly delineated and the latter triumphed. This seemed to allow some members of the FNAL community to externalize or neutralize the present funding and management crisis—the strategic decision to leverage the future of the field with large hadronic colliders.

Initiation Narratives: High-Energy Physics and Individual Empowerment

The tensions that have charted the responses of these physicists to the change from small- to large-scale experimentation reflect the contradictions that framed discussions about the future of the field. The "next machine" has always been the grail of many accelerator physicists. But lobbying for the next machine is inherently risky when weighed against the social and financial costs of such technology. The bigger the machine is, the more distant the relationship between physicists and instrument. The nature of the technology, along with the sheer size and cost of the enterprise, demands large collaborations, both political and scientific. Such collaborations demand the symbolic reordering of the intimate ties between physicists and the technologies of experiment—ties that are central to the identity of many FNAL physicists. As a result boundary rhetorics that address the changing sociopolitical landscape also work to redefine and stabilize the identity of community members.

The mythos of the independent scientist seemed to provide the necessary counterbalance to the anonymity and dubious democracy created by large collaborations. When interviewed, many participants told vivid stories of their initial experiences in physics—stories that resonated strongly with FNAL's ethos within the system of national laboratories. These narratives focused on participants' attraction to the subatomic world and the research technologies required of high-energy physics. They offered accounts of life as an intellectually curious individual. Most significant, these stories were expressive of participants' deeply held sense of accomplishment and uniqueness. By focusing on what made them unique, as individuals and as scientists, FNAL employees reclaimed some of the power and prestige lost in the transition to large colliders and detectors.

I began many of the interviews by asking participants how they came to work at FNAL. Most began their answers with stories of their first exposure and initial attraction to the field of high-energy physics. For some the attraction to particle physics research seemed to grow out of the substance of the research itself—the subatomic world of quarks and leptons. One graduate student commented: "My first semester, I couldn't take a physics class because I didn't have calculus yet. So the physics department had a class for physics majors to get them together, before they could take physics.... The title of the book was *Quark,* and I was hooked. I had never heard of quarks before *Quark*—I was hooked. It was great! I was like, 'Oh wow, this is cool. This is neat. This is what I'll do!'"[19]

Despite years of difference in age and experience, a senior FNAL physicist echoed this same sentiment, saying:

> There's something which is really magical you go through when you're in college and you're learning physics.... At some point it actually dawns on you that everything that goes on around us could be described by a mathematical equation. That's pretty radical, if you really think about it.... The fact that, you know, you could calculate these things that are happening around you and that these equations actually predict behavior and performance of things and that there are underlying rules to the game in the universe and that there's this very beautiful network of laws that kind of balance everything. That appreciation for nature is not held in the general public.[20]

These stories were more than "how I chose my major." They were intense self-revelatory moments accuracy in a rhetorical situation full of memory and emotion. In describing their education, many participants pinpointed "aha" moments when it was clear to them that they saw the world in different terms than "ordinary people" do and that their capacity to do so was somehow strongly connected to physics. For them physics was like a form of second sight, a rare and powerful way to perceive the natural world. These participants told their tale of self-discovery in the same form as they might communicate a moment of scientific discovery; it was as if their realization was not the product of gradually increasing interest and skill, but rather the result of an encounter with some deeper truth about the world itself. Stories of discovering one's essential identity as a physicist seemed to have many of the same formal properties as stories about discovering new particles, perhaps not coincidentally.

Moments of self-discovery were often articulated by way of metaphors that invoked the vision made possible through physics technologies. Many of those interviewed described their experience of excitement and intrigue when they first encountered technology that made it possible to "see" into the subatomic world. Detailing her first encounter with the products of particle-physics research, one participant commented, "That first course I had used some of those old bubble chamber photos. That you could see the track and measure the particle from the curvature—I loved it."[21]

The excitement of particle physics and its technology was also evident in participants' stories about their first encounters with Fermilab. Descriptions of the laboratory often featured the technology as a character equal in importance to the human actors at work. Human action and power were frequently defined in technological terms. Two of those interviewed were strongly influenced by college trips to FNAL. One described her first introduction to the laboratory this way: "We came here for a tour of Fermilab. I remember seeing a bunch of people with flannel shirts in the control room, all these monitors, and looking really busy. One had two phones and I thought, 'That's what I want to do! I want to be in the control room. . . . ' It's kind of like being an air-traffic controller. I thought that was neat, so I took a couple more physics classes in college and ended up here eventually."[22] The other participant's story articulated a less specific and yet somewhat similar theme: "I found out that they were building one of these big labs up near Chicago and I thought that was kind of cool. I came up on a bus trip with a bunch of other undergraduates, and this place was really cool! I decided this is what I wanted to do because it looked so very challenging."[23]

As described in these narratives, the decision was fairly straightforward—the work looked exciting and challenging and so they decided it "was really cool." Notice how these responses are peppered with recurring references to accelerator and detector technology. It appears that the allure of the technology was key, not only to their early identity as physics majors but also to an emerging professional identity. One of the previously quoted participants eventually did get the opportunity to be "in control" and discussed the power and responsibility associated with directing the overnight shift on a small experiment. "We have two people on shifts and you're shift's captain so you have to make decisions about the experiment. If some electronics go bad in the middle of the night, you have to make a decision. 'Shall I try to replace this?' It put you in positions that you might not otherwise get the opportunity to be in, so it's kind of exciting. It's scary, but it's exciting."[24]

The nature of being a physicist was not exclusively defined in terms of technology, however. Many participants also described seemingly innate qualities that defined them as scientists. For example, several described childhood experiences that marked them as unusually curious about the natural world and scientific subject matter, a quality that distinguished them from their peers—for good or for bad—in their early scientific education. As one woman explained: "In science class it was memorizing things, so I went into class and I asked [the teacher] why he wasn't teaching us about electrons. This is what I wanted to learn about and I just demanded it in the middle of class. I got myself into all kinds of trouble. I was not his favorite pupil after that. . . . I was probably not very nice about it. I wanted to know about electrons!"[25]

This physicist noted that in her situation she was doubly vexed—she represented not only a challenge to accepted teaching methods and curricula but also to accepted gender roles. Being marked as different, despite the difficulties associated with the label, was important to this individual and many other participants as well. Often marginalized as children, many seemed to reclaim these early experiences in adulthood by focusing particularly on the selective nature of high-energy physics as a field. One participant reveled in the fact that the culture of high-energy physics encouraged this interpretation of his identity. He commented, "The job I do now . . . there's probably a couple hundred people in the world who can do it, maybe a hundred, maybe fifty, something like that. . . . I wrote [my family and said] 'Your brother is literally one in a hundred million!' Only fifty people on the face of the earth do this, so that's like another version of looking at yourself [as] one in a million, right? It's better than that. And it's true!"[26]

It was clear that these participants identified strongly with the fundamental nature of their research and also enjoyed the power and prestige that high-energy physics seemed to afford them. Equally important, however, was the way in which physics worked to integrate personal and professional identity. When asked how they would present high-energy physics to the public, it is perhaps not surprising that many participants reiterated elements of their personal narratives. Many participants spoke of the importance of conveying the excitement of high-energy physics and sharing their fascination with the subatomic world. When questioned as to how she would describe her work to nonscientists, one participant commented, "It has to sound interesting, and it has to bring out the 'Wow! Cool!' type response that you get when you show them pictures of galaxies or picture from Mars."[27] In response to the same question, another scientist talked about the "next machine" as the collective dream of the community and suggested, "Give talks all over the place about this dream. . . . 'We have a dream. We don't know if it can be done, but this is what we have in mind and this is why we want to do it. . . .' You know, get them excited."[28] A third participant reiterated the best aspects of his job and offered this response, "Oh . . . the passion, excitement, mystery . . . finding out at three in the morning the discovery nobody on the whole planet knows but you . . . things like that."[29]

The desire to share aspects of their professional experiences illustrates the essential tension between identification and division that underlies much boundary work. Anxiety about the problematic evolution of the field and declining public support for the physical sciences finds an outlet in the technological sublime. The excitement and wonder associated with the sublime did not seem contrived to most participants but was instead an obvious choice because it reflected the most significant experiences of their work life. For most of those interviewed, representing these emotions and ideas to the public afforded an opportunity to *communicate,* to share substance, rather than an attempt at persuasion. The desire to share experiences was itself tempered by the acknowledgement of difference.

Despite their positive personal experiences, however, many participants remarked on how the meaning of their work was diluted by constant pursuit of funding (for bigger machines) and the stress of managing large experimental collaborations. Thus, in their attempts to court the public with attractive and exciting images, they also reinforced concepts important within the laboratory's culture and the daily practice of particle-physics research. With complex disputes in play about how best to preserve the future of the field, it was not surprising that the community brought forth a boundary discourse that emphasized the major achievements of the laboratory and also the intimate connection between experimental technologies and skilled users. The images and concepts that shape the rhetoric of the site and the self-tour resonate with the experiences physicists claim brought them to the field—the allure of the subatomic world, the power and control made possible through technology, and the challenge and accomplishment they experienced in the practice of physics. To stand at the rhetorical gate of the laboratory and say, "this is who we are," is an attempt to unify the community behind a commonly held identity and fundamental understanding of what it means to be a FNAL physicist. This boundary discourse, though designed to appeal to the diverse public that the laboratory serves, also functions as a stabilizing force within lab culture—a symbolic axis mundi pointing the community to its almost-religious core values and central identity markers. The act of thinking about how to present the lab and its work to outside audiences had prompted a substantial amount of reflection on the part of the community; but reflection focused communicating community identity and values to any number of "others." The failure of the SSC had also forced physicists to reconsider what held them together as a community in the face of this contemptible "stupidity."[30]

FNAL and Its Diverse Publics

Reflecting on the self, in this case the communal self, necessitates the construction of "others." As a federally funded institution, the laboratory has many of "others" who are constituted as "publics"—those communities whose constituent members or individuals that can and do exercise influence over the laboratory. This section details the community's assumptions about the public and considers how the rhetorical construction of "others" has influenced that laboratory's boundaries.

The problems faced by the physics community have prompted reconsideration of the power and influence of groups in the funding and support of basic research. University fiefdoms and ties to the military-industrial complex had previously insulated the community from outsiders and obviated the need to consider how nonscientists might relate to or make sense of research. As the power to control research was gradually distributed across an ever larger and more complex system in part by political contestation, physicists began to attend more closely to interest groups important to the future of the research. In doing so, they began to formulate assumptions—some accurate, some not—about public interest in particle physics.

Today most physicists recognize that they necessarily communicate with a number of different publics at a number of different levels. In the past, government agencies devoted considerable time to the elucidation of clear-cut and predetermined categories to define the social groups involved in publicly funded research. The labels invoked—scientists; taxpayers; the interested lay public; nonscientists, and so on—illustrate well the desire to create a sorting system of mutually exclusive classifications. Not surprisingly the physics community hoped to simplify this "public zoo" in the same way they had tamed the "particle zoo." When asked about the nature and challenges of "public communication," most participants began by parsing the publics according to reductionist logics not unlike the classifications that govern their scientific work. As one participant described it: "Public support consists of lots of different constituencies. . . . It's our neighbors, it's the general public, it's Congress, it's Department of Energy, the university community, not just high-energy physics people, but the intellectuals in general in universities and we need to communicate with all of these different groups and probably communicate in different ways."[31]

Although this participant's model included a fairly large number of constituencies, several other participants distinguished simply between the general public and the interested public: "There are two publics. There's the general public and I think it's important to at least answer any of their questions. There's the [interested public] . . . they read *Scientific American* and take physics courses in college. I think we could do a lot more with [them], especially when they're sitting in that physics course. Those people actually can understand what we're doing and I think we can have a large impact there."[32] The interested public, a category created in large part by surveys designed to measure public understanding of science, was significant for many participants. It was in many ways a demographic defined according to education and habit—a collection of individual participants credited with both the capacity and motivation to understand science. In short the interested public was thought to be educable about physics. Addressing these sympathetic fellow travelers became important.

The boundary between education and communication was negotiated in a variety of ways. Many participants simply equated the two, but others distinguished between communicating with adults and communicating with children. Communicating with adults was a matter of public relations, whereas communicating with young people was education. Drawing distinctions according to age and mode of address, one participant offered a detailed description of the interests of various audiences:

> What I found talking to the public is that if you talk to a second grader, they ask the kinds of questions that you're interested in. Second graders want to know how fast comets go or what they're made of . . . how long it's going to be until the sun blows up and what happens when you get to the end of the universe. These are the kinds of questions that scientists care about. Teenagers ask about careers and how much money you make. You can answer those kinds of questions, but they're not interesting. . . . Adults want to know how much money you're spending and how many foreigners there are at your lab or about space aliens or some really complicated question about wormholes.[33]

This individual was clearly most comfortable with second graders and inferred that other scientists shared her feelings. For her, publics were organized according to the level of their interest, age, and development. Her talk implied that people were pretty much washed up by the time they reached adulthood, having lost their innate curiosity about the natural world and become preoccupied instead with the mundane details of the everyday life. Furthermore, in this view adults lacked the capacity to change their opinions or learn new ideas. Communicating with "like" publics was enjoyable because identification emerged from the exchange of sublime experience and the purity of intellectual curiosity. Children's questions resonated with this participant's values and experiences as a scientist, particularly as they revealed the complexity of seemingly simple ideas and preference for nature's wonders over the messiness of politics. In contrast, when asked to talk with adults at the laboratory's open house, a number of her like-minded colleagues simply hid.

Some participants were more comfortable with adult audiences. When describing encounters with visitors who asked questions about expenditures at the laboratory, many participants used the term "taxpayer." As one participant reminded me, "This is all paid for by the taxpayer and there's no reason to hide it; we're proud of it."[34] Another argued, "It's their money. We are obliged to tell them what we are doing with their money; tell them we're not wasting it."[35] A former administrator talked in terms of the public's ownership of the lab, "We're doing all this for the public. This is the public's; they own it!"[36] These comments point to the peculiar isolation of "taxpayer" as a category separate and apart from "the interested public." A large percentage of those interviewed defined taxpayers as the ultimate source of funding and therefore marked them as an important public in their own right. The designation "taxpayer" carried certain performative expectations and implied a clear relationship to the laboratory and its employees. Taxpayers funded research but should not have expected to benefit directly from it. They could inquire about how their taxes were spent, but they could not contribute to decision making. Most physicists felt an obligation to be able to explain the nature and purpose of their work to the taxpaying public if called on to do so, but they did not regularly seek any significant input or involvement from nonscientists. Emphasis was placed consistently on *recognition* and *acknowledgment*, rather than *involvement*, of the public in the life of the laboratory. As one participant described: "They're paying for this . . . I figure out what my budget is and, given my budget, there's probably twenty to thirty families out there in this country whose taxes pay for my research. That's just my personal research budget, not to mention the Fermilab utility bill."[37]

Clearly this physicist is sincere in her appreciation of the public funding, working to personalize her understanding of "the public" by thinking in terms of "families" and their tax bills. And, yet, these families are still abstract, existing at significant remove from the laboratory. While entitled to acknowledgment, they are not seen as decision makers in the funding process. As one participant explained:" The public is one thing, the decision makers are another. I think sometimes the public is more enlightened than the decision makers are. Once upon a time, the people who knew no science at all were discarded from society—now they're elected to Congress!"[38] Other participants reinforced this distinction between the public and those who make decisions that affect the laboratory. "Well, there's a difference between informing policy makers and informing the public," one said. "That's where the money comes from; not from talking to people here in the war zone."[39]

Many participants singled out government officials and members of Congress for blame in the current funding crises. Participants commonly expressed concern about the motives, intelligence, and educational level of elected officials. Some spoke of the "revenge of the C students" and sometimes labeled DOE and other government officials as "failed scientists."[40] Others were much more charitable and claimed that "congressmen and senators are incredibly sharp" and employ aides that "really listen . . . [and] really try to understand what's going on."[41] Despite some positive comments, however, most participants felt it was particularly difficult to convince government officials of the necessity of international collaboration in large-scale physics research. Recollections of the demise of the SSC prompted comments such as the following: "With the SSC there was a combination of insisting on foreign involvement, while at the same time there were congressmen saying, 'We don't want these fuzzy little foreigners getting all of our secrets.' There were actually congressmen who were against it because they'd learn our technological secrets."[42]

While many criticized individual members of Congress for being insufficiently informed and appreciative of science, others traced the problem back to a faulty educational system that did not prepare elected representatives to make such decisions. As one scientist explained: "There's probably some guys that are even intelligent, but it all goes back to the schools. . . . The schools are bad so the people that you elect have gone through these schools and, in comparison to what I see in the European community, I think their level of sophistication is rather low."[43]

Overall, then, those interviewed classified the lab's public according to age, role, interest, and decision-making power. From these comments, three key audiences emerged: the "learners," comprising primarily children, college students, and interested adults; the decision makers, including governments officials, members of Congress, and administrators at the DOE; and the general public, an abstractly defined category populated primarily by taxpayers. Conspicuously absent were specific comments about those living in surrounding communities. These individuals, when discussed, seemed to be included with the general public.

Participants freely discussed the nature and quality of the relationships they had with each group. Most felt comfortable with children or college students, audiences they had been exposed to in the normal course of their work. Building on this experience, many defined "interest" as a product of quality education and thus envisioned the interested public as graduates of an effective system of science education. Many participants focused on the importance of the third audience, the decision makers, noting that this portion of the public had the most impact on the future of the field. Consequently this audience seemed to command the majority of the community's time, energy, and effort. The general public, often equated with taxpayers, seemed to be the least clearly defined and therefore the least understood of all the audiences. Although many participants discussed the importance of being accountable to taxpayers, the limited understanding of this group undermined most efforts to communicate with them.

These audiences were constituted by the physics community in an effort to map the territory of the laboratory and the field of high-energy physics. Not only do participants' comments draw distinctions between scientists and nonscientists, they organize the world of nonscience into three groups, each with a different set of power relationships and communicative norms. Learners are constituted within a didactic frame. Dialogue is possible with this audience, but only within the prescribed power relationship between teacher and learner. Decision-makers, in contrast, are constituted within an adversarial frame. They are regarded as resistant to the needs and values of the community. The general public is addressed primarily through monologs that emphasize acknowledgment rather than communication.

Perceptions of the General Public

Physicists in the FNAL community harbor a number of problematic assumptions about the general public that work to reinforce boundaries, heightening the experience of difference and distance and diminishing the possibility for identification and shared understanding. The belief that the public is less educated and more self-interested than the scientific community and therefore less qualified to make decisions about the future of the field was common among those interviewed. Many participants acknowledged the public's right to be involved on some level, but almost all erected boundaries on the grounds of education, critical-thinking skills, and understanding of the process of scientific research. The role of self-interest was more complex. Although a few participants regarded the public as entirely self-interested, others argued that the public might be genuinely curious about physics but lacks adequate information that might prompt an enthusiastic support for high-energy physics.

Most participants argued that the general public they sometimes address is largely undereducated about physics and, more broadly, about basic principles of scientific method. They expressed further worry that such a lack of scientific grounding left audiences vulnerable to unscrupulous, unscientific claims. One participant commented extensively on the "crisis of rationality" he felt was evident in American culture: "When you see almost a million men in Washington keeping their promise or whatever they're doing, you worry. I don't mind people being religious, that's fine, but this fundamentalist notion is so far away from the goals of public understanding of science and a commitment to rationality. That's the crucial point. What we represent here is the commitment to rationality, which is why we have a technological civilization. It's a commitment that started way back with Galileo and it's too late to turn back. You may be happier trusting prayer to cure your diseases, but the track record for prayer is not very good and the track record for science is very good."[44]

For this participant, the concern is not simply that people might vest their faith in something that could be personally dangerous to them but also that fundamentalist religion (and other social institutions) threatened the authority of science in American culture. Many other participants echoed this concern, often citing specific examples that they felt illustrated a lack of critical reasoning ability that results from or is exacerbated by deficient science education. One scientist commented: "When I was teaching inner city high school kids, one thing that struck me was the level at which they didn't understand the world. It was a tremendous revelation to them that light switches were connected to lights by wires. If your understanding of the world is that far removed, I don't see how you can tell the difference between . . . a pretty amazing thing like that and aliens landing on Mars. I mean, which is more fantastic?"[45]

Another participant argued: "When I drive here I go by a Holiday Inn [where] every other week they have a psychic fair. You have astrology sections in the newspaper—there's more astrology coverage than science coverage. I grew up in an area where people don't know the difference between astronomy and astrology. People would rather listen to the Farmer's Almanac than the National Weather Service. I'm sort of used to that mindset."[46] Not surprisingly, critical thinking and rationality figured heavily in the comments of most of those interviewed. These skills were often described as "qualities" or "capacities" that defined what it means to be "scientific," and therefore they were seen to be prerequisites for participation in decision making about the future of scientific research.

Many participants critiqued the public's capacity for critical thinking through examples that dealt specifically with the role of science in risk analysis. Undoubtedly concerned with the public's ability to assess the risk of living next to (or on top of) a large particle accelerator, these participants cited examples in which popular assessment of risk erred—situations in which physicists correctly understood a particular risk while the public did not. One participant spoke at length about his view of fallacious causal reasoning on the part of the public: "Cell phones are a perfect example. I mean every physicist knows, after thinking about it for ten seconds, that there's no way that a cell phone can cause tumors.... [Also] this controversy about overhead high tension [lines] ... I mean they can't possibly cause a tumor. You need much higher-energy waves to cause a genetic change, but because there were random fluctuations and one of those fluctuations happened to be under a power line then 'power lines cause them.'"[47]

These examples were offered as proof that the public's judgment can and should be questioned in cases of risk assessment. Many of those interviewed were particularly skeptical of public calls for precautions in scientific research and decision making. Troubled that government policies and regulations are often guided by public opinion, one scientist commented: "People don't understand that what you know today may be different than what you know tomorrow and so science knowledge changes and your scientific understanding changes.... If you see that something is dangerous, your reaction is to prohibit it. Saccharine was that way, right? It was supposed to cause an enormous amount of cancer. As you understand [these things] more, sometimes they become worse, sometimes they become less worse. It changes so you have to get that story across to people."[48]

In theoretical terms, the restrictions this individual envisions represent boundary incursions—the unauthorized and unwelcome intrusion of public opinion into scientific practice and decision making. Many participants felt that excessive attention to public opinion could lead to unnecessary restrictions on laboratory operations and several cited the aggressive "Tiger team" inspections by the DOE in the 1980s as a case in point. Others interviewed, however, felt the scientific community played a significant role in the public's perception of risk. These participants argued that it was the responsibility of the scientific community to convey information about risk clearly and in a timely manner. Citing an accidental tritium release at Brookhaven National Laboratory that had caused a public outcry in Long Island, one scientist commented: "You've got to be sensitive.... This person doesn't want to drink as much tritium as a physicist does. Those poor people around there. They never received any physical harm, but their property values have been horribly hurt and they've gone through a lot of stress and anguish, worrying because they don't understand. To them, any level of radiation is dangerous and they have no real reason to suddenly start believing these guys."[49]

For this individual and others like her, incidents such as the tritium leak at BNL provide crucial opportunities to model rationality and build public trust in the national laboratory system. Somewhat surprisingly, few participants talked in detail about building trust with Fermilab's immediate neighbors. Several indicated that they knew people who lived in the surrounding towns and felt the laboratory had a good reputation in the community, and yet few offered specific examples of interaction.

Despite seemingly few interactions, lack of public trust was a persistent topic of conversation in many of the interviews. Many participants pointed to the legacy of the military-industrial complex, speculating that the public had a very limited understanding of the distinction between physics research and nuclear weapons work. In short they worried that the public might think weapons production takes place at FNAL. As one participant commented, "Nobody knows what goes on at Fermilab. They still think we're making bombs here. I think you need to show them that it's still a dynamic field . . . and that most physicists aren't building nuclear weapons or working on some new defense system."[50] Participants also hoped to avoid associations with nuclear power. One participant who had once worked at the lab's reception desk commented, "Most people around here believe we have something to do with making bombs or maybe we make our own power, like nuclear power."[51]

Secrecy was a primary concern of many participants. As the previous chapters detail, the laboratory's ethos was founded on the appearance of transparency and openness, a rhetoric intended to thwart associations with nuclear weapons or power. The installation of a guardhouse and control gates at the laboratory entrance seemed to undermine this effort. One participant worried the gate would serve only to fuel public suspicions about the lab's mission: "We used to open the walls to the public around the area; now we've closed it up. Even at the time that we had the lab open to the general public for traffic, they still were thinking that we're building bombs or we're doing something secret here. Now we've closed it up; they think even more that we're doing something secret."[52] Echoing this sentiment, another respondent argued for the importance of "telling people we don't make bombs here."[53]

Lab employees' sensitivity to confused perceptions of research, nuclear, and weapons labs may also stem from their own varied but largely negative experiences with the military-industrial complex. As one participant commented: "When Admiral Watkins was in charge [of the DOE] it was the absolute pits because he used to give orders . . . and we don't take orders. He imposed a mentality which was suitable for nuclear submarines, that's where he came from. It was a good idea in terms of trying to clean up those reactor messes, which were really a disaster, I mean people were sloppy. We're not in the same ballpark. We couldn't do that if we tried. So they tried to do that to us, all these OSHA and DOE regulations came down. . . . All they did was grind operations to a halt."⁵⁴

This participant drew a sharp distinction between military and research installations. He questioned the degree of oversight deemed necessary at the research labs, claiming that the type and degree of contamination seen at other national laboratories could never occur at Fermilab. Refuting the very possibility of contamination, he argued further that preventative oversight is actually a barrier to good research. Similar arguments have been offered to explain management problems at the SSC, where military-industrial contractors and managers were hired to work side by side with the high-energy physicists building the accelerator: "[President George H. W.] Bush took a lot of guys who were in charge of weapons programs that were shutting down and put them there so it was the first machine that wasn't built really well. The physicists weren't in charge, there were lots of conflicts, and it was a big mess. . . . As a community, we were quite humiliated."⁵⁵

Feelings of humiliation reverberate throughout participant's stories of the SSC. The involvement of military managers and defense contractors in the project threatened the autonomy that high-energy physics had fostered since the mid 1960s. Sociologically the success of the high-energy physics labs was measured by the capacity to appear to operate separate and apart from the rest of the system. The conflict was experienced as a clash between two very different organizational cultures: "There were a lot of problems with oversight and management . . . having the physicist's culture interact with the military-industrial complex, which had a lot of control in the system.⁵⁶

As the community members focus on the next machine to be built, they are brought back to the political problem of the SSC. Some participants responded by reinforcing the boundaries around the laboratory. One suggested the project be presented to the public as "a Fermilab project" in order to forestall both negative public perceptions about the military-industrial complex and problems with managerial control. Others recognized that divorcing the project from associations with the military-industrial complex might be a double-edged sword. A few laboratory employees challenged the notion that the public had a universally negative perception of the military-industrial complex and argued that, since a portion of the public tends to regard defense work as more useful than basic research, the lab may have actually benefited from the public's confusion in years past: "The fact is that's where they get their money from . . . that slight confusion on the public's part: 'Gee, don't you make nuclear power over there?' 'Don't you have some bombs and stuff like that?' "⁵⁷ Regardless of the direction of response, however, it became clear that participants forged the identity of the field in relationship to the military-industrial complex, but rarely addressed their discourse to audiences within that culture. Instead it was "the public" that was the audience for boundary negotiation at Fermilab. Participants' responses were most often formulated with assumptions about public attitudes or perception in mind. Their comments differentiating the laboratory from the military-industrial complex were almost uniformly addressed to a general audience.

From Education to Public Relations

Seeking refuge in academic culture, many participants characterized the confusion of science and the military as a problem of misperception and lack of education. With this issue and others, communication was most frequently defined in terms of the relationship between an educator and idealized "student-citizens." Reiterating themes common to early literature on the public understanding of science, many participants argued that improved training in scientific method in the early grades would lead to a citizenry that demonstrates the proper respect for organized research and its contributions to society. Many also advocated improving the college curriculum for both science majors and nonmajors alike. Lastly most participants defined museums, open houses, and lab tours as sites for informal education and discussed the need to improve these sites as well. In contrast to their staunch stance on the precise explanation of scientific principles and methods, almost all participants agreed that scientific education, both formal and informal, must be delivered in an exciting way that would motivate learners to take an interest in larger issues facing science and scientific research.

A number of participants proposed specific, content-based solutions that would create an educated public. One scientist commented on the problems of the standard college physics curriculum: "We have a very stuffy old curriculum where you basically learn the nineteenth century, and I think that's our own damn fault."[58] Another noted the mundane nature of many informal educational displays and argued, "I firmly believe that we've got to do a better job telling the general public what we do and why it's exciting, why it's interesting."[59]

In contrast some participants pointed out that it was not their early education that led them to science but some pivotal experience or direct encounter with the practice of scientific research. As a result experiential or interactive learning vehicles were often offered as an alternative to traditional, lecture-based methods of instruction. One researcher speculated: "We could be reaching very large numbers of people by including high-energy physics examples in basic physics [courses] because people really do think it's cool. If you have one quarter of physics, it's cool, and you can understand a lot. . . . I like to spend seven weeks teaching them electromagnetism, and then we go to Fermilab, and I show them how we actually use all this stuff."[60]

Another participant emphasized the importance of starting early and stressing the relationship between basic research and everyday life experiences: "Number one, people have to be educated from day one. Number two, they have to see how this stuff integrates our basic research and is part of their experience and part of what's important."[61]

Creating excitement was central to most curricular proposals, formal or informal, mentioned by participants. One described how she would reorganize the tour of her facility for the next open house: "We brought everyone in and had these posters set up so we gave everyone a small physics lecture. . . . They kept looking at the detector, and we probably should have shown that first. We should start out with the exciting stuff, so they could sort of sit there and look at it while we were talking."[62]

A strong relationship between excitement and the sublime was also apparent in her description of how she taught open-house visitors about the basic elements of her experiment: "I was showing them, [saying], 'this piece of steel is our detector, and it blows up tiny little protons.' I was like, 'How cool is that!?' I said, 'You know about molecules, you know about atoms, and those are all neat . . . let's go even smaller and deeper and don't you wonder what those are made of and how that works?'"[63] Examples of similar teaching techniques were repeated across the interviews.

It was clear that most participants felt fairly confident of their ability to educate the public, perhaps for that very reason; however, the creation of education programs to serve that purpose was also a source of some controversy. The majority of participants advocated some form of direct participation in both formal and informal public science education, but the motivation for becoming involved often differed according to age and experience. For some older physicists participation was the best way to exercise control over the material presented. These individuals often advocated a top-down model of educational control in which curriculum would be developed by working scientists and disseminated to the public via traditional science education. As with the issue of laboratory management, education appeared to be a space in which competing groups struggled for institutional power. For many younger scientists, involvement in education was a means to demonstrate their willingness to communicate with those outside their community in hopes of improving public support for research. These physicists, whose career paths had changed significantly since the cancellation of the SSC, advocated a system that stressed informal education tailored to the needs and interests of the audience rather than the demands of old-style scientific literacy. A member of the public affairs staff elaborated on the origins of this generational difference: "A full generation of physicists were getting paid, quite well in many cases, to do research into fascinating subjects because the country thought that it was just nifty.... The younger generation know that if they want to do particle physics, they're going to have to convince people who have the money that it's worth doing. Some do it better than others, but it's not something they feel resentful about or think they shouldn't have to do, which the old guys do."[64]

All respondents agreed that doing nothing to improve public understanding of particle physics reflected badly on the community and the public's perception of their respect for the taxpayers who fund their work. Commenting on what he perceived to be the poor quality of older exhibits in the self-tour, one participant said, "I have gone up there, and I can explain it pretty well, but most people will not know what the heck they're looking at. ... I think [it implies] that the scientists don't care what the public says. They do what they want to do, and they don't give anything [back]."[65]

Discussions such as these often called into question the boundaries between education and simple public relations. Though many lab employees conflated public relations and education, the creation of the public affairs department and the experience at a recent open house prompted several participants to consider the distinctions between the two. A number of those interviewed distinguished between their desire to participate in educational programs and their ability to serve as spokespersons for the laboratory. While most felt comfortable in various formal venues ranging from the college classroom to a second-grade field trip, they often lacked confidence in their ability to make their case to the general public, particularly adults. One scientist commented: "I think it's more the norm that when a reporter calls up a physicist, he or she thinks, 'Oh, my god!' It's not part of our education—we're not trained to communicate. We sit in the classroom, someone talks at us for an hour, we write in our little notebooks, we go home and solve problems and hand [them] back. There's no verbal communication whatsoever in the course of the entire education system."[66]

Another participant elaborated on the problem by drawing a sharp distinction between education and entertainment, teaching and persuading:

> We have a public affairs office that works on our image and I think that's reasonable. I think scientists enjoy and actually should be part of the process of educating the public, but they are often a mismatch. From a public image standpoint, last weekend is a great example—the "universe in a jar." The universe in a jar is a jar. You put corn syrup in it, put a little sparkle in it and a label on the top and you're out of there. If your purpose is to make everyone feel good about having come to an open house at Fermilab, and feel good about having a lab in the neighborhood, this is a good project. Most of the scientists at the lab heard

about it and said, "What's the point?" "What are you learning with that?" "What are you teaching with that?" There's a difference between educating and entertaining and doing work on your image. This is more advertising than educating. This is the kind of thing that you would do at a trade show.[67]

Many participants thus simultaneously supported efforts at increased public communication and yet tried to limit their personal participation in such work, given the way it was being institutionalized. The motivation for drawing a boundary between the "educating-self" and the "advocating-self" differed among participants. For some it was a matter of skill. For others such work was somehow undignified or outside the ethical framework or social contract of science. As one scientist argued: "It's hard for scientists to deal with that kind of issue. And I don't think it's really a scientist's job.... The set of skills that are that level of wheeling and dealing should be divorced from each other. I don't think they should be intertwined."[68]

That participant felt most comfortable leaving the "wheeling and dealing" to specialists. Many others agreed that the use of professionals could increase the effectiveness of the lab's message. For example, one scientist made the following suggestion:

> What I would really like to see is the science community put money into advertising. If you look at the AMA, they do incredibly well. Everybody tells me, "Go out and talk to your congressmen." I think it's a hell of a lot better if you get these professionals. You can laugh at it, and everybody looks down on them and stuff like that, but the advertising people get the message across.... If they'd charge every scientist a hundred dollars a year and put it into a fund [to] go after publicity in a professional way, I think you'd see a completely different atmosphere.[69]

Another participant agreed and argued that even reports to funding agencies should be "glossy," "well constructed," and "written by someone who is a science writer."[70]

The comments in this section reveal more than physicists' attitudes about science education and public outreach. They are a window opening onto the scientific community's perceptions of the general public. In most cases lab employees regarded the public as undereducated and lacking in critical thinking skills. This depiction of the public resulted in a curious articulation of the boundaries between education and persuasion. Many participants emphasized the power of education to correct what they saw as a "crisis of rationality" that might threaten the future of the laboratory. In doing so, they clearly recognized the persuasive power of the education and often talked of education as a means to promote the laboratory and its interests. Others, noting the work of the public affairs office, drew a strong distinction between education and public relations. These individuals preferred to leave the work of promotion to the professionals and in consequence often removed themselves from contact with the general public altogether, including education.

Arguing for Basic Research

Despite their disagreement over the appropriate role of scientists in the *presentation* of arguments for basic research, participants were remarkably consistent with respect to the *content* of those arguments. They discussed three appeals that could be used to garner support—patriotism, technological spin-off, and knowledge for knowledge's sake. Not surprisingly, each of these themes can be traced back to the discourse that first shaped the creation of the national laboratories. Patriotism is a remnant of the cold war. Technological spin-off speaks to lingering critiques of the value of investing in basic research. Knowledge for knowledge's sake is the counterbalance to spin-off, a distant cousin to the earliest arguments for the importance of basic research.

At FNAL patriotism was described as "the pride you get in your country having these facilities."[71] Discussing the impending start-up of the large hadron collector and America's impending loss of the high-energy frontier, one participant commented: "I think if people know enough about what we do now and you present this as the highest-energy particle collider in the world and [explain] that it will be gone soon—that seems to ring home with them, [the idea] that Americans are the best at this but, we're not going to be for long. Would you want to let that slip away?"[72]

Such comments were tempered, however, by the acknowledgment that the present course of physics research too is dependent on international collaboration. There was a recurring tension between nationalist discourse and talk of international cooperation. The idea of international cooperation is not new, and in fact some of the most symbolically charged attempts at engaging it were made during the depths of the cold war in the form of the world accelerator movement and the Pugwash conference, which since 1957 has brought together scientists, often from opposing nations, to reverse the negative uses of their own discoveries and inventions. The Fermilab accelerator itself thus became an object useful for working through the competing discourse of nationalism and globalism. It was open to the world: "I think a lot of people like to see an American accelerator as 'all ours' and it's possible, but there's a lot of foreign collaborators that come here and research, and we would be lost without them."[73]

The public affairs staff acknowledges that "the days of doing these things as a single nation, one nation doing it alone, are gone."[74] But they also admit to the lingering effectiveness of nationalistic arguments with some audiences "We talk out of both sides of Fermilab, depending on our audience. We have twenty-three hundred users, of whom eighteen hundred are U.S. users, and the [others] are from foreign institutions. Depending on who we're talking to, we either emphasize that we have these eighteen hundred U.S. users and forget about the other ones, or we can talk about what a great melting pot it is and how we have people from twenty countries. We're very selective about what we emphasize there."[75]

The end of the cold war and the changing structure of large-scale research initially suggested that the outlets for these patriotic appeals would decrease in number. Although participation in projects such as the large hadron collider will no doubt be discussed in patriotic terms, it is unlikely that anyone will think of it as an "American" project. The majority of patriotic appeals are now focused on the machine that will follow the LHC and the prospect of bringing the energy frontier back to the United States. If the next machine becomes viable, promoters will undoubtedly face the same questions that plagued the SSC. Should the project emphasize the realities of international collaboration and thus invite outside involvement, or should it present the machine as America's reclamation of the energy frontier? The latter approach has a cathartic appeal for those once involved with the SSC. As one participant joked, the next American machine could "show the LHC [at CERN] up for the pathetic second-run, second-rate machine that it is!"[76]

Deploying patriotic arguments will always be problematic, however, so many in the physics community have focused instead on other arguments in favor of supporting basic research, including the link between basic research and the development of new technology. When promoting the SSC, some physicists claimed credit for everything from superconducting materials to magnetic resonance imaging. This strategy backfired dramatically when materials scientists disputed the community's claims, shifted the argument frame, and made several very public appeals for more government funding for applied rather than basic research. The rhetorical failures of the SSC did little to change the physics community's beliefs, however. Although many will acknowledge that mistakes were made in public debate about the SSC, most still believe that there is a fundamental link between basic research and technological and economic growth. These beliefs give way to two dominant argument strategies intended to demonstrate the value of the laboratory to the larger culture. The radical "spin-off" argument asserts that there is a direct connection between high-energy research and the development of technologies in other areas. A more conservative "economic power" argument contends that basic research drives economic growth, but it makes no claims with respect to specific technologies.

Strong spin-off arguments are far less prevalent today than they once were. When asked, most participants acknowledged that attempts to credit high-energy physics with technological innovations usually associated with other fields has damaged the community's reputation and relationship with other scientific and engineering communities. Some physicists, however, still offer examples intended to demonstrate a clear and direct line between basic research and practical technology. For example, one participant argued: "There are a lot of things that you don't realize that probably wouldn't exist if it weren't for high energy physics, like computers. We've pushed computers to their absolute limit . . . so we feed the computer industry. We give them money—that's why they build better computers, and that's why you're able to buy a relatively cheap PC that works fast."[77]

The link to computing is becoming an increasingly common example intended to demonstrate the usefulness of physics research. During my time at the laboratory, a public internet-access computer was installed as part of the self-tour in order to demonstrate the connection between physics and the development of the World Wide Web. Speaking at the height of the "dot-com" boom, accelerator physicist Rich Orr credited physics with the genesis of electronic commerce: "I think what we do here is the basis for *everything else that's going on*, even the basis for the '.com' [that] corporations are putting after their names to make billions. The World Wide Web was invented by high-energy physics. It's important that we keep doing basic research. It's important that Fermilab stays alive."[78]

Here credit is distributed broadly across the high-energy physics community with only extremely indirect references to CERN, the initial institutional home for the World Wide Web. The discourse also fails to reference the many innovations in programming that allowed the World Wide Web to develop to its present state. This rhetoric represents the Web, a complex technological innovation with diverse contributors, as the whole-cloth creation of high-energy physics.

More recently physicists have revived older arguments that emphasize the development and use of superconducting materials, the strategy that backfired in the case of the SSC. Reacting to public concern about energy prices and the status of the electric grid, advocates for a very large hadronic collider have made public claims that basic physics research can lead to the development of superconducting transmission lines. Such rhetoric often acknowledges that such technologies are not yet possible, and yet graphic images of power lines and particle accelerators underscore the association and collapse the barriers between idea and reality.

Such blatant attempts to argue for spin-off are decreasing, however, as the field faces criticism from fellow scientists in applied physics. As a result many high-energy physicists have modified their strong claims about the technological outcomes of their work. Conservative arguments focus on abstract concepts such as economic growth and job creation. One participant addressed the problematic nature of the term "spin off" as a label for what he saw as a variety of outcomes that may result from basic research: "It's actually not spin-off; it's actually economic power this country gets from the repercussions of those inventions. Economists know that half the GNP is paid back in technological innovation—costs dropping, getting more functionality for less money, medical expenses going down. Diagnostic testing is completely different now because of all of these inventions. That's why I think spin-off is a bad term."[79]

This participant still made strong claims about the role of high-energy physics in the development of more practical technologies, but his manner of presentation focused on the abstract relationship between the gross national product (GNP) and technological development rather than concrete connections between experimental technologies and practical technologies. Building the connection in this way discouraged questions or counterarguments about specific adaptations of technology for practical use and the consequent role of scientists in applied research, engineers, and industry. The claim made here was in effect similar or larger in scope than that of its more radical cousin, and yet it appeared more conservative by virtue of its lack of specificity.

Conservative arguments about technological pay-offs were very common in the literature produced by the public affairs office. Convinced that such appeals were persuasive to general audiences, one public affairs staff member explained the approach: "The message is [that] basic science is important for our country. Our country needs to be among the leaders in basic science. There's a long time between discoveries in basic science, . . . but that connection does exist."[80]

Once again the emphasis is on a strong yet abstract connection between basic science and technological and economic outcomes. The goal of the conservative approach seems to be to reinforce the idea that such a connection exists, while assiduously avoiding both specific claims that might prompt challenges from applied scientists and engineers. It seemed equally important to avoid specific predictions about what new technologies might result from ongoing research, lest the community be held accountable for some measurable result or benefit. We see echoes of nationalism in such talk, as the support for physics and its institutions is coded as a measure of patriotism.

Some of those interviewed, however, speculated that either form of the spin-off argument was problematic, in part because it placed the community at odds with its own motivation and goals. One participant worried that inordinate emphasis on the technology of particle physics might lead the community away from its central focus and consequently mislead the public: "What [high-energy physicists] are doing is [answering] the fundamental question, 'what's it all made of?' I don't see the spin-off from that. What you have to do to answer that question—like build this magnet, build that detector, build this communications system—those things apply, but what you're after is an intellectual exercise."[81]

Several participants worried that it was insincere to promise, even indirectly, technological outcomes where none may result. These individuals, while well versed in the common spin-off rhetoric, felt such indeterminate outcomes demanded a consistent emphasis on scientific rather than technological objectives. As one participant put it "[there] is indirect spin off, but you cannot go out and sell it to the public [by saying] because of this indirect benefit, we have to do this scientific project."[82]

The majority of those interviewed rejected spin-off arguments even though they were well versed in such rhetoric and equipped to use it if called on to do so (for example, at the lab's open house or in a meeting with government representatives). As our discussions progressed, most participants made their way back to more basic arguments in support of high-energy research that were closely aligned to their own motivations and beliefs about the role of fundamental knowledge in the larger culture. These knowledge-for-knowledge's-sake arguments have a long history in the field of high-energy physics. One can find them prominently displayed in Lederman's *The God Particle*.[83] They are also key to the rhetoric of the self-tour, where knowledge and rational understanding of the sublime are presented as a goal unto itself. The physicists interviewed offer a somewhat paradoxical explanation of the usefulness of these appeals. On one hand, some participants admitted that they felt the majority of the public was not swayed by the knowledge-for-knowledge's-sake approach. One scientist commented: "I think it's a smaller group that you can sell the idea of knowledge-for-knowledge's-sake [by stating] 'it's exciting!' Who cares about the top quark or things like that? You're going to have a harder time selling that kind of thing to a legislature that has to give you money."[84]

When asked, however, to describe the social importance of their work and to identify outcomes that make high-energy physics worth doing, most resorted to just such "knowledge" justifications. As one participant explained: "What we do here is answer a pretty simple question [such as], 'What's it all made of?' At least that's how I try to impress it. [You ask the question and] you carry it to its ultimate end."[85]

Another described how his attempts to understand nonscientists' perspectives actually led him to emphasize basic knowledge in his communication with the public: "What I try to do is view this laboratory from the perspective I had when I was back home . . . or the perspective my parents might have or my in-laws when they come to visit. [I emphasize] very basic questions like, 'Where does mass come from.' I ask kids sometimes, 'Do you know why the sky is blue,' [or] 'Why does the sun look red when it's setting?' These are basic things that are happening around you all the time, yet most people don't sit and question it."[86]

Although these participants and others felt that the answer to such questions would lead ultimately to more practical applications, they believed the knowledge gained was sufficient justification in itself. Echoing the sublime rhetoric of the self-tour, one participant had in fact argued: "Like the pyramids, [high-energy physics] is driven by the desire to understand what our existence is about and how it came about. Without any immediate benefits to society except possibly spin-off, [it's] mostly the spiritual [we] try to understand—why are we here and how did we get here?"[87]

For others the research process was imagined through a peculiar conflation of childlike wonder and philosophical reductionism. These participants emphasized the importance of experimental particle physics to the production of scientific knowledge and self-knowledge. The idea that physics research is fundamental to both science and human experience permeated a number of interviews. One participant expressed the same idea in personal terms: "As a kid you go out and watch the stars and wonder where they are and what would happen when you get to the end of them. It's gotten so that we know what things are made of, not perfectly, but to a level that goes beyond what most people think about. There's so much we don't know about the stars that people do think about . . . that grabs the public's attention."[88]

This participant drew heavily on the sublime as she discussed common experiences of staring in awe and wonder at the night sky. Her belief that such questions of origin, substance, and function are truly universal provided her with a conduit to a public from which she felt otherwise distanced. Positioning physics at the center of both human experience and inquiry seemed to provide the community with a fallback framework less prone to the problems of either patriotic nationalist or spin-off arguments. While the patriotic or technological achievements claimed by the physics community can and have been challenged by powerful outsiders, the realm of basic questions, or knowledge for knowledge's sake, has always been the territory of the basic sciences. In order to challenge such claims, one would need to disentangle the personal motivations of scientists from the concerns of the publics to be served by this work. Knowledge for knowledge's sake appeals to our desire for self-knowledge, and as such the value of high-energy physics is a personal rather than public judgment. The success of these appeals is dependent on the construction of identification between scientist and nonscientist. Such arguments resonated strongly with the experiences of those interviewed, and it may be this grounding in experience and the desire for identification that sustains this rhetorical strategy in an age when pragmatic appeals are increasingly popular and necessary.

Although the internal logic of the physics community suggest the origin and function of knowledge-for-knowledge's-sake arguments, there has been little exploration of the public's response to this or any of the other rhetorics considered here. Accordingly the following chapter explores the nonscientist's perceptions of the relationship between science and society and tests the physics community's assumptions about its audiences through a detailed analysis of the laboratory's public exhibits and texts and visitors' responses to these texts.

FIVE

Stakeholders, Self-Tours, and Communication after the SSC

As we proceed down the stairs, I feel like Alice descending into the rabbit hole. Even though I know the tunnel is partially underground, this descent seems significant. All my views of the accelerator have thus far been from aboveground, often fifteen stories aboveground as I view it from the self-tour gallery. The tunnel environment is dark . . . and relatively deserted. In sharp contrast to the openness of the aboveground facilities and the site itself, the tunnel features chain gates that block access to the accelerator (when it is running). A number of other spaces are similarly secured by lock and key. But the most notable thing is the dirt . . . and the rust. This building contains the world's most powerful particle accelerator—a machine capable of circulating protons and antiprotons at speeds that defy imagination—a machine that symbolizes the essence of the modern technoscience. Aboveground one envisions it as the stuff of **Star Trek** *episodes—the pulsing glory of the plasma core, an ultraclean conduit for nearly a terra electron volt of electricity. But it is dirty. Not just dirty, it is filthy dirty. There is a huge contrast between this machine and the Cockcroft-Walton machine—the accelerator that starts the cascade and is the frequent star of Fermilab's public and media tours. The Cockcroft-Walton is a gleaming tribute to modernist design in indigo blue and stainless steel. In comparison the bigger and younger machine, the Tevatron, appears very old indeed.*

   I am confident that the disappointment of dirt will be replaced by the pleasures of scale. I imagine that my walk around the ring will provide me with the experience foreshadowed in the lab's videos and displays. But we have scarcely entered the tunnel when we must stop and turn around. Our guide's Geiger counter is rattling with increasing intensity, and a familiar swath of yellow and black tape shrouds the next section of the machine. We are standing about thirty feet from a hot spot, a place where the beam went astray and contaminated a section of the tunnel. Once again the reality of the situation strikes me, and the invisible purity of the machine is sullied. Radiation reads like a betrayal of the sublime promise of the accelerator as it is viewed from above, a decent into the terrifying that is the sublime's other side. The spell is broken at the same time, but the machine also becomes more real in some very important ways. It is just what it is—a four-mile string of magnets and wire encased in a massive cryogenic cocoon. It breaks and needs repairs. It is fallible. It is dangerous under certain circumstances. And, in the words of our guide, "Some days it's amazing that it works at all."

With the emergence of "stakeholder" publics after the SSC cancellation, audiences identified in terms of their investment in the success or failure of research programs and institutions began a new chapter in the rhetoric of the national laboratory system. Despite previous efforts to preserve the authority of the physics community and insulate it from outside interference, those opposed to elite governance in high-energy physics seized important rhetorical ground in the early nineties, when they began to question the necessity of large-scale facilities devoted to basic research. When combined with previous failures to communicate effectively with nonscientists, the emergence of a powerful, historically grounded counter-rhetoric thrust the lab into a new set of social relationships that challenged the secure boundaries of the laboratory. Since the late eighties, and continuing to the present day, the "public" has become an increasingly persistent and influential voice in the future of high-energy physics. The failure of the SSC not only signaled the community's inability to make its case to Congress but also the importance of public opinion and communication in the decision-making process. As the controversy over SSC smoldered during the eighties and early nineties, members of Congress and representatives from competing branches of science took their case against the facility to the media. The emerging rhetoric echoed the conflicting ideologies that characterized the debates of the forties, fifties, and sixties. While the controversy was fresh, its rhetoric mapped familiar territory.

As they had so many times before, physicists misjudged the power and privilege they enjoyed in relationships with influential senators and representatives. They wrongly assumed that proscience alliances were forged through the involved parties' shared beliefs about the value of basic physics research more than through the opinions of their constituents. These same constituents took on new force and importance in the wake of the SSC debate. Because the battle for the SSC had been waged through congressional hearings and mass-mediated texts, a once abstract and poorly understood public suddenly became concretized as a participant in the policy-making process; it was one among several stakeholder groups who could make claims on the future of the laboratory. By placing the public alongside the lab's faculty, users, and staff, the DOE and other government agencies sent a clear indication that the national laboratories must attend to public concerns or risk the consequences that might result from challenging a stakeholder's claim.

Stakeholders, Taxpayers, and Public Affairs

As earlier chapters have shown in more detail, the history of physics funding is peppered with references to the public and their reasonable expectations regarding the value and outcomes of investing in a large-scale program of basic research. Many representatives and senators argued, as they do now, that the public should expect some technological payoff for their "investment" in the laboratory. When the cost overruns of the SSC made national news, the benefits and drawbacks of basic research were once again thrust into the national spotlight. Those opposed to the facility had a history of pragmatic arguments to draw on, while the physics community had few tangible payoffs to which they could point and claim ownership.[1] Nonetheless Fermilab's staff attempted to redefine its social role within the framework suggested by the stakeholder relationship. The stakeholder public now had power underscored by institutional practice. Laboratories across the national system were now accountable to citizen-stakeholders in ways they had never been before in these discursive conditions. Revelations about environmental contamination and safety hazards spurred many of the most significant changes, including the creation of community advisory boards to that ensure a mechanism was put in place for the nontechnical public to voice their view. Many of these changes were not fully implemented at labs that did not have a history of contamination or crisis. Yet boundary work across the system now had to focus intently on the relationship between the lab and the newly empowered public. As a result, much of the rhetoric produced was focused on the need to define and limit the claim of most stakeholders. Furthermore, with questions being raised about the necessity and worth of basic research, high-energy physics labs increasingly focused on distinguishing themselves from other competing branches of basic research in which the political public might choose to invest.

John People, the laboratory's third director, came to this situation early, assuming leadership of the lab in 1989. Although he had spent most of his career at FNAL, People brought a unique sensibility to the role of director. His background included many technological projects and intimate connections with SSC. He was appointed director of FNAL in 1989. But in 1993 he was appointed to supervise the shutdown of the SSC and in fact served as director of both facilities simultaneously for a nine-month period. People's undergraduate degree in engineering and his work experience at Morton Aircraft distinguished him from other physicists in his cohort, although not always in a favorable way. He had neither the drama and elite sensibility of Wilson nor the reputation and affability of Lederman, but he did have a track record for getting difficult jobs done on time and, when necessary, rescuing programs in trouble. His management style included a great deal of delegation. Sensing that he was not as skilled a communicator as his predecessor, People instituted a public-affairs office within the laboratory in 1995. Rather than entrust the lab's rhetoric to outsiders, the in-house public-affairs office was staffed in large part by individuals already working in the laboratory or by communication experts who possessed significant knowledge of or connection to the world of particle physics. The office made a point, however, to hire professional science writers to produce stories for the media. Most important, in the wake of the SSC, the public-affairs office was expected to work in collaboration with the lab's scientists to manage the FNAL's image in Washington and other important communities.

Recent changes in how physics research set the agenda for the new public-affairs office. One scientist commented that the lab "was more than just having a place the public could come to, *you had to be more aggressive* [and] vigorously compete for a little bit of time with congressmen."[2] Whereas the earlier public-information office had been largely reactive, the new public-affairs office was to be proactive in the hope that the lab might anticipate and plan to address emerging public concerns. People emphasized that the office should break with the past and work to serve public needs. The old public-information office was conceived of as a resource for the public, yet it had no obligation to seek out the lab's constituents or even manage controversy. In contrast the public-affairs office was modeled with a more aggressive outreach program in mind—one that included soliciting news coverage and actively shaping the lab's image with the press. To this end the newly created office came to manage the laboratory's public texts, instituting a systematic refurbishing and upgrading of public exhibits and designing or redesigning the various brochures and pamphlets that were made available to visitors. The lab's Web site was upgraded and expanded to appeal to nonscientists as well as scientists. Finally, the lab's newsletter (*Ferminews*) was modified so that it could serve both the laboratory community and interested outsiders.

Organizational change centralized the outreach efforts of the public-affairs office under the watchful eye of the director. The old public-information office had been part of the laboratory services department and was thus well removed from the most influential aspects of laboratory life. Locating the new public affairs office within the directorate, the seat of power in the laboratory, signaled its importance and made clear the effort and resources to be directed toward communicating the lab's mission to the public. Despite these lofty goals and ready access to both funding and institutional power, however, change was fairly modest. The nature and long history of the laboratory's deeply embedded sublime narrative worked against a complete reevaluation and reformulation of existing rhetoric. Long-standing perceptions about elite governance and the impact of public opinion persisted, and, as one participant related, the justification for change was vague at best: "You come into a job like this with a set of notions that are about the last war. . . . There's not enough time that we can sit back and think of how we want to change the course. There's certainly been an evolution, but I would say that it's been driven by the observation that we need to be more and that the public is asking us for more, or at least expecting more."[3]

Though the desire to change was strong, the lab's rhetorical goals were poorly defined. Even as they acknowledged the public's desires and concerns, the majority of those involved in the new efforts still assumed that the important audience could be found in the halls of government. They supported the idea of active management of the lab's image for outside audiences, but their efforts remained focused in large part on persuading influential members of the federal government and decision makers in the relevant government funding agencies. As a result the strategy developed by the public-affairs office recycled themes, ideas, and images from the past. The constraints of scientific practice, the unique and strongly held sense of identity, and an inability to reassess the actual rhetorical situation they were in resulted in a program that worked to reinvigorate the rhetoric of the sublime, rearticulate the importance of the laboratory and its technology, and build on the cultural authority of the experimenters who work there.

The Self-Tour

The influence of the public-affairs office was felt most strongly in the reconceptualization of the spaces and the texts of the laboratory, particularly the self-tour. The self-tour introduced visitors to the mission and work of the lab. It was here that the reiteration of the existing themes to which I have just referred was most strongly developed.

Refurbishing the fifteenth floor of Wilson Hall and the exhibits associated with the self-tour was one of the principal concerns of the newly formed public-affairs office. To do this, it enlisted the help of lab employees with experience or interest in this type of work. The resulting displays on the fifteenth floor are some of the few "public texts" of the laboratory that have emerged from the combined efforts of individuals representing public affairs, education, and the scientific staff of the laboratory. While the development of the self-tour was officially overseen by public affairs, the effort was led in part by Ernie Malamud, staff scientist and founder of SciTech, a science and technology museum in neighboring Aurora, Illinois.

With assistance from a consulting firm in Chicago, the committee attempted to clarify its message to visitors. The principal goal, as articulated by those involved, was to reassert the identity of the laboratory and its staff ("who we are") and introduce visitors to its work ("what we do"). Emphasis was placed on upgrading exhibits to reflect a coherent, professional sensibility. Contemporary media production techniques gradually replaced what some in the public-affairs office considered to be outmoded methods. In this sense the self-tour—more so than other texts produced with the aid of the newly inaugurated office—became a venue through which the laboratory struggled to define the social meaning and value of its work. In the following section of the chapter, I detail the rhetoric of the self-tour, particularly the ways in which the tour constructs visitors' understanding of the accelerator and shapes their perception of the community of scientists who work at the lab.

Sublime Rhetoric and the Boundaries of the Laboratory

The self-tour is structured both to give visitors a sense of the extremes of scale that characterize particle-physics research and to mark its technology and "products" as sublime. This construction of the sublime is further reinforced by oppositions that mark FNAL as a sacred space, most notably through the contrast between primitive and developed (overdeveloped) landscapes and natural and technological ways of relating to the world. These oppositions can trigger a range of emotional responses—awe, wonder, and confusion—as they point to the presence of the sublime. But evoking the sublime for rhetorical effect is tricky business. Writings detailing the nature of sublime experience suggest that such far-reaching and powerful combinations of emotions cannot be sustained for long periods. For example, one of the emotions evoked by sublime experience is fear, be it from actual danger or inability to comprehend the object or phenomena contemplated. The subject must resolve the conflicting emotions evoked on the one hand by rational understanding or domestication of the sublime object and on the other by the fear that impels one simply to withdraw to a place of safety.

The need to encourage sublime experience translates into a set of discourses, images, and experiences that define FNAL's relationship with various publics. Of chief importance is the *permeability* of the boundary. Permeable boundaries allow for a wide variety of relationships that can change as the circumstances of the ecosystem demand. Dense, impermeable boundaries work to fix identities and power relationships. In rhetorical terms an impermeable boundary rhetoric functions to make identities essential and remind audiences of their place in the social structure. It also manifests power through self-reference and draws attention to gaps in understanding and shared meaning that exist because of the boundary.

In this sense the rhetorical boundaries drawn around FNAL unite those within the community by differentiating scientific understanding from nonscientific understanding. The technological sublime manifest in the FNAL self-tour can be resolved only through a very specific interpretive framework (for example, an understanding of quantum mechanics and the nature of high-energy physics), and thus the boundary drawn constitutes audiences in terms of their capacity to find specific meanings in a sublime object. As a result those sympathetic to the habits of mind of the physics community are the most likely to transgress the boundary. Those unable to "think like a physicist" may be forced to retreat and accept the power differential that the boundary implies. In this sense the confrontation with sublime phenomena during the self-tour functions as a test or rite of initiation. The boundary between scientist and nonscientist is articulated through the ability to interpret and understand the exhibits in terms recognizable to the FNAL community. This is an extremely important and often unpersuasive aspect of FNALS's presentation of itself in the self-tour. Inevitably visitors, even those who aspire to cross the boundary and "become members of the community," must depend on the scientists to explain the experience. Visitors' identities are defined in terms of their distinctive responses to the sublime.[4]

Prior to 11 September 2001, the self-tour was intended to provide visitors with a very specific, scripted experience of the laboratory, its science, and its technology. While free to wander the grounds as they liked, visitors' activities within Wilson Hall were tightly controlled. They were directed to areas usually devoid of actual scientists or scientific activity. The notable exception is the cafeteria, although few of the visitors I interviewed understood that they were allowed to eat there. Once on the fifteenth floor, the focal point of the self-tour, visitors were positioned as passive receivers of information and oriented to four sources of "information" about the laboratory: static exhibits; models; panoramic views of the accelerator and grounds; and films. While most exhibits involved a great deal of reading, listening, and complex interpretation, there was no one to whom visitors could pose questions about the scientific and technological information they encountered. In that sense, it was truly a "self-tour"—an articulation of the self in relationship to the laboratory, its technology, and the meaning of science in contemporary society.

Static Exhibits

The majority of the content of the pre-9/11 self-tour was communicated through immobile exhibits and films. A variety of static exhibits and "disembodied" accelerator technologies were on display for the purposes of communicating the laboratory's identity, the role of the experimenter in contemporary scientific research, and the relationships between the lab and its overseers in the Department of Energy. These exhibits labored to celebrate the relationship between the "tools"—the accelerator and detectors—and their skilled users. Furthermore the relationship among the exhibits underscored visitors' experiences of both the mathematical and the dynamic sublime. Images of the laboratory grounds directed visitors' attention to the extreme size of the accelerator and the vast natural landscape. Depictions of the interior spaces of the accelerator and detectors heightened their awareness of vast regions of "inner space" and reiterated the power released through particle collisions. The exhibits repeatedly juxtaposed the size of the particles against the extreme power required to "produce" them. Thus the sense of scale produced through the static exhibits extended along two axes: one of size and one of power—and the smallness of the individual.

The various static exhibits can be also categorized by topic: exhibits that explained the mission and administrative structure of the laboratory; exhibits that described the people at the laboratory; and exhibits that explained aspects or concepts from particle physics research. The majority of displays dealing with the mission, administration, or people of the laboratory were clustered in the north end of the self-tour area—in the section described by a lab employee as "who we are." Exhibits with content about research were found in the south end of the display area—the section devoted to "what we do." There were noticeable and somewhat inexplicable exceptions to this logic, but, generally speaking, the first section of the tour established the identity of the laboratory and its research community, whereas the second section reinforced an interpretation of "accelerator as crucible."[5] In what follows I describe the exhibits in the present tense as a visitor would have experienced them before the post-9/11 changes.

The following two examples are typical of the first two types of displays. The first illustration, entitled "Fermilab Serves University Scientists," communicates the number and national origin of those who work at the lab. The most dominant features in the exhibit are two photographs of specific collaborations that demonstrate the large numbers of scientists who use the lab. Yet sheer numbers are not the only issue referenced in this exhibit. The text, which reads "the majority of U.S. high-energy physicists use Fermilab's Tevatron . . . ," encourages readers to assume that the lab is the preeminent facility in this country or alternatively that there are no other high-energy facilities in the United States. The ambiguous use of the word "majority" fails to acknowledge the large number of researchers working at Stanford Linear Accelerator Center, at Brookhaven National Laboratory, or even at CERN, and, in doing so, it seems designed to reinforce FNAL's identity as *the* National Accelerator Laboratory. The text also enumerates "thirty-two states" that are represented by the lab's personnel. An emphasis on national identity is clearly communicated by the second-most-dominant image in the exhibit—a map of the United States, including Hawaii. Scientists from countries other than the United States are mentioned only briefly in the text and represented by a small footnote near the bottom of the map. In this sense both the images and the text reflect competing narratives of international collaboration and nationalism that characterize much of the public rhetoric in high-energy physics. The post-SSC environment demanded that a choice be made between identifying FNAL with international collaborations that were—and are—arguably necessary to ensure the future of "the next machine," or distinguishing the facility from its competitors in the United States, Europe, and Japan. This exhibit seemed to suggest that when confronting the unexpected vacuum in institutional leadership that followed the cancellation of the SSC, FNAL chose to reinforce its identity as *the* premier American facility for high-energy research.

Fig. 8, entitled "Science at Fermilab," focuses further on the distinctiveness of Fermilab and its mission. Once again images and text argue that "science at Fermilab" is somehow different from science in other places. A view of the lab grounds occupies the top portion of the picture. The image of Wilson Hall and the Tevatron ring takes up the center third of the display. The track of the new main injector anchors the bottom third of the display. Superimposed over the base photograph are two more images—one a computer representation of a particle collision and the other a photograph of Wilson Hall and the Obelisk. The computer image represents the sublime world of the very small. The background picture reflects the very large. By choosing this particular computer image, the small is connected to the large through the common arc of the circle. In the small photo, Wilson Hall is reflected in the pond in the foreground, forming an image that extends both upward and downward from the horizon, suggesting that structure is somehow rooted in the earth. Wilson Hall is thus depicted as the nexus between the seen and unseen, the superterrestrial above and the subterranean below, and, most important, the gateway to the accelerator.

The Obelisk in the foreground is similarly reflected. For its creator, Robert Wilson, it represented the necessity of individual courage and creativity in the midst of collective efforts to achieve greatness. As it stands alone in front of the building that houses the collective leadership of the laboratory, it suggests the synergistic relationship between the individual and the collective in scientific work. Central to the exhibit is the text, a reprinting of FNAL's mission statement, "Fermilab advances the understanding of the fundamental nature of matter and energy by providing leadership and resources for qualified researchers to conduct basic research at the frontiers of high-energy physics."[6] This statement focuses not only on the sublime task of high-energy physics—"understanding the *fundamental nature of matter and energy*"—but also on the agency required to accomplish that task—"providing leadership and resources for qualified researchers." The words "leadership and

Fig. 8. "Science at Fermilab"

resources" are superimposed over the more distant image of Wilson Hall and the accelerator cascade implying that these represent the resources necessary to take physicists to "the frontiers of high-energy physics." The accelerator therefore is no ordinary instrument, but rather one that can probe previously unknown regions of inner space. These images of the laboratory recur throughout the self-tour and the literature of the lab. The public-affairs office has abandoned many of the pen-and-ink illustrations that epitomized the lab's image in the sixties, seventies, and eighties. Yet the new designs, emphasizing professional photography and computer-designed layout, are strikingly similar to the earliest depictions of the lab that were hand-drawn by Angela Gonzalez. The clear connections between the two, despite dramatic differences in media, testify that much of the laboratory's rhetoric is grounded deeply and inalterably in highly specific ways of looking at the lab and its technology.

Figs. 9, 10, and 11 represent the third type of display—exhibits detailing specific aspects of particle physics research. Fig. 9, entitled "Collision!," places emphasis on the power and energy required for and produced by a particle collision. Particle collision is central. The balance of the image suggests both the symmetry and productivity of the collision. The letters and symbols identifying the various particles also indicate the diversity of the products produced by collisions. Bright colors—yellow, red, and orange—suggest energy and power. The overall effect illustrates the paradoxical nature of the sublime. This collision is both destructive and productive. Another display in this series compares a particle collision to a car crash in which two crashing Volkswagens might produce, for example, two trucks and a race car. For most viewers the result obviously is somewhat unexpected and defies everyday understanding of the outcomes of a collision, which are are commonly understood as destructive. This inference, in combination with a visual emphasis on symmetry, supports a degree of magic as well as a degree of control of the result not generally associated with powerful collisions. Presenting the image in these terms suggests that productivity and order underlie even things that initially appear chaotic.

Fig. 10, entitled "Is It a Top Quark?," attempts to demonstrate how the top quark "signature" is identified in computerized representations of particle interactions. The language in the display is generally technical (calorimeter, detector, cell) and devoid of explanatory context. The images and visual organization of the display are somewhat similar to fig. 11. The subject of the exhibit, a computerized representation of a particle collision, is centered on the page. As the viewer moves through the display, the image is "unrolled" to reveal a "Lego plot" of the resulting collision events. The Lego stacks are identified for the viewer as jets, but no further explanation or interpretation is offered. Given this decided lack of explanatory context, the viewer may be left wondering whether this is a top quark or not. The point of the exhibit seems less to explain the interpretation of collision events than to preserve the viewer's perception that expert interpretation is required to understand the processes of the subatomic world.

Similar evidence of mystification can be found in fig. 11, a cutaway illustration of a superconducting magnet. Just as the previous display obscures rather

128 The Boundaries of the New Frontier

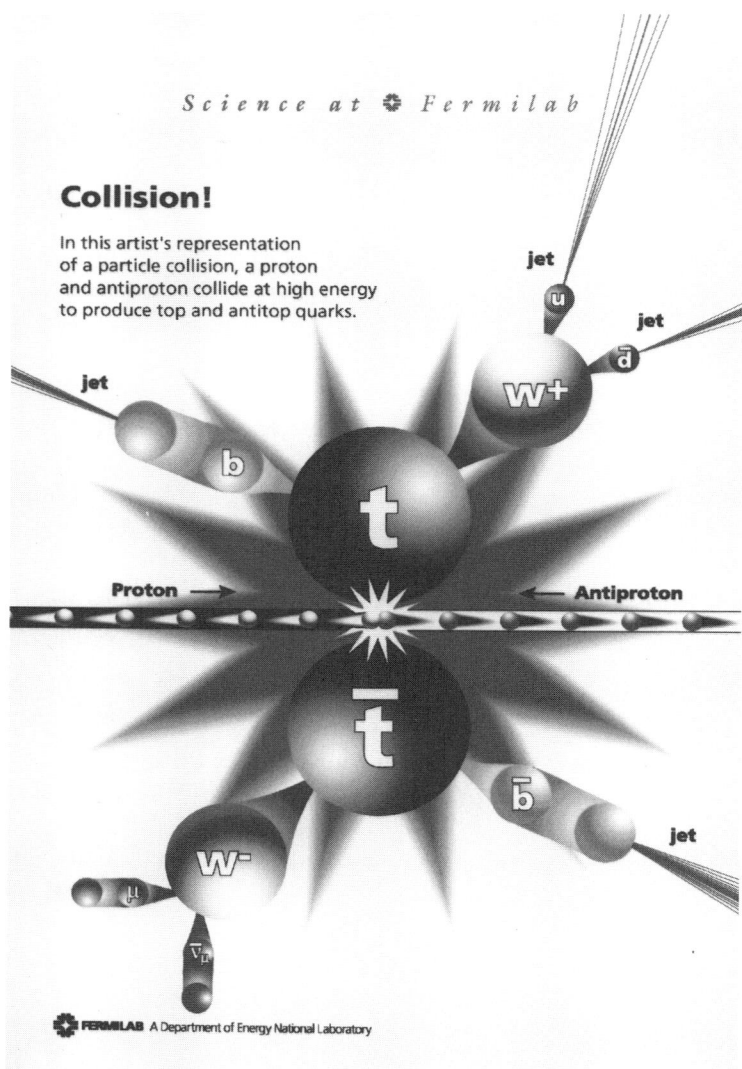

FIG. 9. "Collision"

than explains the process of interpreting a collision event, this exhibit mystifies the interpretation of one of the central technologies used in such research. Although various parts of the magnet are ambiguously identified in this illustration, there are no explanations as to how the magnet works, nor are there any intertextual references that would point the viewer toward a more detailed understanding of magnet technology.

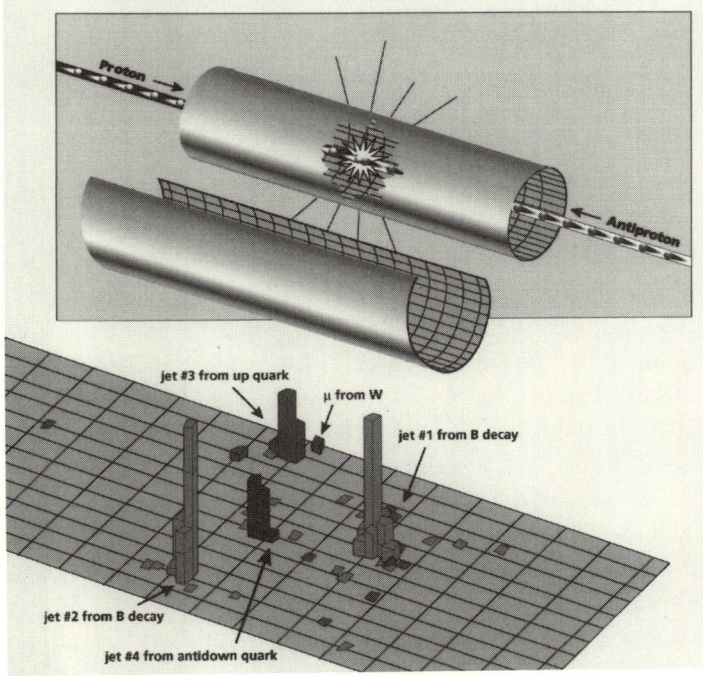

Fig. 10. "Is It a Top Quark?"

In addition to static displays, the self-tour includes various "disembodied" technological components, many displayed with no interpretation or explanation. Within the larger context of the self-tour, these can be read as technological relics, meant to provide visitors a glimpse inside the accelerator and detectors. When considered in relation to other exhibits that emphasize the extreme power and energy of particle collisions, these "altars" encourage

130   The Boundaries of the New Frontier

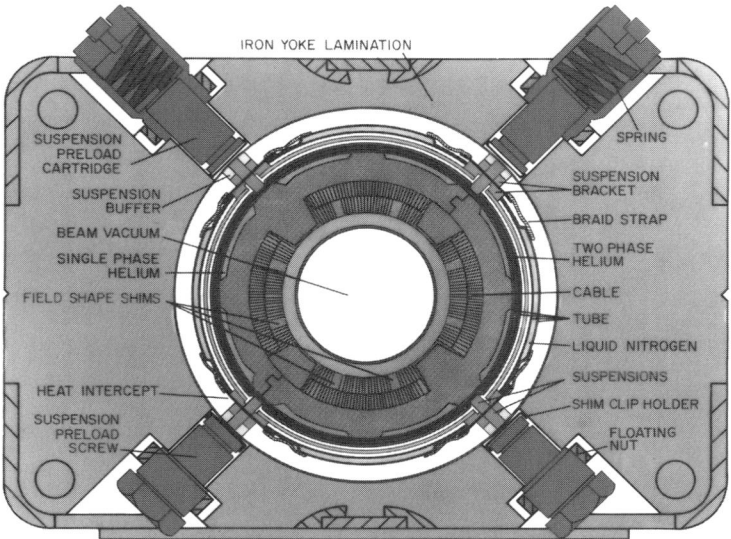

Fig. 11. A cutaway illustration of a superconducting magnet

viewers to marvel at the sublime nature of the machine itself. In many cases the emphasis on the aesthetic experience of the machine is in no way accidental. Several of the displays were originally developed for an art exhibition entitled "Virtue: Homage to Physics Artisans." The exhibit presented elements of physics technology as art objects, and thus these pieces were not originally designed with the same pedagogical or rhetorical motivations as other aspects of the self-tour. However, when embedded within the self-tour, the exhibits seem to encourage the viewer to contemplate the paradoxes and oppositions that emerge from the dynamic sublime—the inexplicable relationships between order and chaos, beauty and destructive power, the very large and the very small.

The illustrations discussed here are representative of the exhibits found throughout the self-tour area. On the whole, exhibits ranged from fairly approachable combinations of image and text that relied heavily on intertextual references and worked to reinforce themes clearly and interpretive frameworks developed in other aspects of the tour to isolated technological altars that simply reasserted the technological sublime amid various attempts to explain it. The modularity of the tour exhibits are of particular importance. Over the course of many years, the self-tour changed in response to changes in the field, the system, and the world outside the laboratory. The modularity of the exhibits, however, allowed for greater stability in the self-tour narrative.

Despite periodic changes in individual exhibits, the stocks of knowledge referenced remained the same. The exhibits consistently referenced key aspects of both the identity and experience of those in the laboratory. Individual displays triggered narratives that underscored the overarching mythos of the laboratory. Thus, though intended to educate outsiders about the work of the laboratory, the static exhibits were ultimately more significant as a mirror that reflected and reinforced the identity of those inside the lab.

Models

Let us suppose that visitors are still able to take the self-tour. They see another major component of the self-tour, the tunnel model, which links the two areas that house the majority of static displays. The tunnel model contains a wooden replica of the accelerator. The display is flanked by mirrors in order to create the illusion of a continuous ring of magnets, thus giving the viewer the sensation of being inside the tunnel. The model is located opposite a long bank of windows that overlook the accelerator and afford visitors the opportunity to compare and contrast the interior and exterior of the accelerator, to imagine what it must be like inside the machine they are gazing down on. The exhibit seems designed primarily to simulate the tunnel experience rather than to explain the design of the technology. Despite recent upgrades designed to improve the verisimilitude of the display, there are no detailed explanations of accelerator components or their functions. This interpretative gap seems intended to focus visitors' attention on the visual elements of the experience and the sensation of being inside the tunnel as a physicist might be.

A large-scale model of the Fermilab site dominates the northern area of the self-tour. This miniaturized three-dimensional "map" of the laboratory allows visitors to organize visually the exterior spaces and, more important, emphasizes the relationship between the various technological elements of the lab. The site model allows visitors to "see" the chain of accelerators, the relationship between Wilson Hall and the Tevatron, and the interactions among the accelerator and the various detector facilities. Much more than a "you are here" type of display (although the model does make it clear that the visitor is "here" at the hub of the lab's activities), the site model permits visitors a limited mastery of the sublime space they have encountered outside. An apron around the base of the display highlights significant features and thus directs the viewers' attention to the central systems of the accelerator and important facilities on the laboratory grounds.

Constructed Views

This process of mastering the relationships among various visual elements of the self-tour continues if the visitor progresses around the fifteenth floor and takes in several highly structured views of the laboratory grounds. The bank of windows opposite the tunnel model is one such constructed view. Were it not for an expressed desire to encourage viewing of the accelerator and surrounding grounds, the self-tour could originate on any floor of Wilson Hall. The main visitors' area was located on this floor in the hope that it would create an appreciation for the scope of the laboratory, its technology, and by extension its work. The exterior views provide the necessary link between the dynamic and the mathematical sublime, inner and outer space, and the visible and the subatomic worlds.

Three views dominate the visitor's experience. Looking south visitors see the circular figures of the Möbius Strip sculpture and auditorium in the foreground linked to the booster ring, antiproton source, and ultimately the main injector. The resulting effect is a chain of circles that gradually increase in size as the eye moves toward the horizon. The eastern view, described earlier, encompasses the entire accelerator ring and on clear days the Chicago skyline. This perspective emphasizes the scale of the ring as the accelerator fills the viewer's visual field. The ring frames its interior space, one of the few areas marked "restricted" on most maps and legends, and suggests this is the most natural or primitive area on the lab's grounds—a pristine site surrounded and protected by the technology of the accelerator. The northern view highlights the fixed target sites and various other experimental areas, many of which are no longer in use. Visitors also see the Feynman Computing Center, the magnet research and testing facility, CDF, and on occasion the buffalo herd grazing in the pastures beyond the magnet research facility. Visible in the foreground are a row of flags from many nations, the reflecting pond, and the Obelisk sculpture. In many ways this is the most cluttered view and requires the most active interpretation on the part of the visitor/viewer.

Films

Visitors to the fifteenth floor could also view a number of videos about the laboratory, its technology, and its work. Two videos that detail the relationship between the laboratory and its publics were available for public viewing at the time this research was conducted: *Pursuit of the Fundamental / Welcome to Fermilab* and *Building Our Future*. A new video, *A Sense of Scale*, has since replaced these two. *A Sense of Scale* updates and includes the recent history of the laboratory, but it references many of the same themes developed in the older films.[7] *Pursuit of the Fundamental* reinforces a number of the themes developed through other elements of the self-tour. The film opens with a series of low-angle shots of Robert Wilson's sculptures and also Wilson Hall reflected in the pond. Following this series of images, the scene shifts to the interior of Wilson Hall, where we see John People consulting with another scientist. People introduces viewers to the themes that will be covered in the video, noting that "Fermilab is home to more than 1,000 scientists [studying] the properties of quarks and leptons."[8] The remainder of the introduction highlights the principal themes of the video: that the lab uses "the most sophisticated technology that we know of" to help "us understand how the universe evolved from the moment of creation"; and that "this technology will ultimately have some use in society."[9] People's introduction clearly indicates the program will focus on the technoscience of particle physics, but he is careful to caution the audience not to "lose sight of" the natural beauty of the laboratory environment.[10]

This comment foreshadows the structure of the text that is to follow. Images of the natural areas, cultural events, and public gatherings are offered in between segments that feature the technology, science, and practitioners of particle physics at FNAL. This segmented structure serves aesthetic, thematic, and rhetorical goals. First, it marks a change in topic and allows the viewers a visual and aural respite (there is no narration during these segments, only music) during which they can absorb the scientific material from the previous section. Second, the interludes give the lab a chance to demonstrate the breadth of activities that constitute life at the laboratory. Third, by associating images of the public with images of nonscientific activities (concerts, nature walks, and so on) these scenes subtly articulate the boundary between scientist and nonscientist. The segments imply that the public can attend cultural events, use the lab grounds for various ecological and recreational activities, and participate in educational programs, but they have only a supportive role in scientific research. Obviously one would not expect to see images of nonscientists participating in the actual work of the laboratory, but it is significant that the scientific segments of the video have virtually no images of scientists interacting with members of the community outside of the context of formal lectures or educational programs. Because there is no connection made between the nonscientific and scientific realms, the lab is treated more like a community recreational and cultural resource than a taxpayer investment. Like learning-based vacations or spiritual retreats, the lab is presented as a place to escape from the stress of everyday life and contemplate greater things. Nowhere does the video mention the lab's funding sources or encourage the audience to contemplate the scope or nature of taxpayer investment.

The bulk of the content is interspersed among the visual/aural interludes. These segments feature discussions about the lab's work, technology, and personnel. The first of these focuses on the lab's mission and the search for a single fundamental force. Subsequent segments feature the cascade of accelerators, fixed-target experiments, colliding beam experiments, detectors, and, finally, computing. In an apparent effort to demonstrate the lab's commitment to human rights and also to distinguish FNAL from stereotypical understandings of scientific institutions, those featured in the film seem to have been chosen to reflect ethnic and gender diversity.

The film makes substantial claims with respect to the capabilities of the lab's technology and the scope of its work. Early on, the laboratory is described as a "testing ground for all of physics reality." This statement reflects a subtle synecdoche that characterizes much of the lab's boundary work. Rather than attempting to articulate the specific difference between high-energy physics and competing branches of physics such as astrophysics or plasma physics, the video simply announces that high-energy physics is experimentally relevant to all. As a synecdoche, such a claim operates on two levels. Not only is high-energy physics substituted for the entire field of physics knowledge, but basic research, defined as the search for fundamental particles/forces, is substituted for the applied branches of research.

According to this film and *A Sense of Scale*, the mission of the laboratory is to produce tools that can facilitate the "continual march deeper and deeper inside the atom and fundamental forces."[11] To illustrate how this is done in particle physics, the video attempts to build a bridge between familiar images of scientific instruments and the more-difficult-to-understand idea of a particle accelerator. The accelerator is compared to a microscope that facilitates vision in a realm where "the nature of seeing changes."[12] In constructing this metaphor, the video attempts to shift viewers' thinking from "waves" to "particles"—particles that when accelerated to high energies and collided can reveal the inner structure of matter. By encouraging visitors to consider "vision" in terms of "energy" and "particles," it becomes easier to equate enhanced vision with increased energy and thus imply that the machine and the work of high-energy physics share a common teleology and a common fate. In order for high-energy physics to accomplish its goals and reveal the fundamental nature and function of matter, the accelerator must be made bigger.

In *Pursuit of the Fundamental*, this common purpose and destiny is clarified through descriptions and images of the existing cascade of accelerators. Viewers are taken through the process of acceleration beginning with the Cockcroft-Walton machine, extending through the linear accelerator (Linac) and the booster ring, and finally into the main ring and the Tevatron. Each stage of acceleration is described in terms of a percentage of the speed of light and accompanied by long shots of the interior of accelerator or tunnel and graphic overlays that suggest the increasing speed of particles as they move through the cascade. At one point viewers themselves "accelerate" through the tunnel as though they were a subatomic particle. Toward the end of this segment, a physicist explains the relationship between the increasing speed of the particle and the increasing size of the accelerators in the cascade. As he describes how the strength of the magnets used determine the bend of the particle beam as it speeds around the accelerator, he asserts that "eventually you must make the ring larger because you won't be able to make the magnets any stronger."[13]

Throughout early segments of *Pursuit of the Fundamental*, viewers see rational explanations of the accelerator as tool that are juxtaposed against sublime images of the accelerator. The narration and explanation offered by featured FNAL staff seem to give viewers a means to "domesticate" the sublime technology by regarding it in terms of a commonly understood scientific instrument, the microscope. The visual images, on the other hand, reinforce the oppositions that define the sublime—the paradoxical relationship between the very large and the very small, the simultaneous existence of matter as particle and wave, and ultimately the extreme energy required to produce a "single" subatomic particle.[14] As the name implies, similar comparisons between the very large and the very small are central to the rhetoric of *A Sense of Scale*.

References to the technological sublime permeate both texts. In *Pursuit of the Fundamental*, one engineer describes the demands placed on the materials in the E-786 detector and jokes that the experimenters needed a "material that was infinitely strong and took up no space."[15] As the engineer describes this, we see images of the actual detector, suggesting that, while we know the material described cannot exist, those building the machine must have developed an adequate substitute. This segment implies that to capture the phenomena being studied, the detector must take on the same sublime characteristics as the phenomena it captures. Although the accelerator ring might be considered sublime by virtue of its size and power alone, the detectors achieve sublimity through their ability to interact with subatomic particles, their capacity to become part of the moment of collision. The featured engineer speaks of "bringing the machine to life." It is here, in the marriage of nature and machine—and the use of machines to bring about material change—that the lines between creation and destruction of particles begin to blur. On one hand the accelerator exists to annihilate protons and antiprotons through repeated high-energy collisions. Such collisions, however, "produce" new particles and facilitate the "creation of new forms of matter."[16]

This paradoxical relationship between creation and destruction is the essence of the crucible metaphor and the cultural narrative of physics to which it points. The technological object does not simply register or observe the collision events; it is an active and integral part of them. The instrument exerts a force on the materials, changing their constituent nature. The consequent "creation of particles that haven't existed since the time of the big bang" is made possible through the intricate relationship of nature and machine.[17] Without the accelerator the collision could not be produced. Without the detector the resulting collisions could not be "seen" (for they only become "visible" by virtue of their interaction with the materials in the detector).

The relationship extends to including the computing power needed to record and analyze the events produced. The viewer sees banks of computers and tape-recording devices, symbols of the almost incomprehensible amount of data generated with each run of the accelerator. These images further reinforce that high-energy physics research is productive of the sublime in all respects. The descriptions of the technological components and the various stages of the research process link the natural and technological sublime through the act of producing new particles. The accelerator is the gateway to the strange and mysterious world of inner space, the realm of subatomic particles and forces. Access to this world, however, is not an end in itself—both the *product* and the *process* are sublime. Although the goal of research is to resolve a paradox of the natural sublime—to reveal the simplicity that underlies the seemingly complex laws of matter—the process whereby that goal is achieved is productive of the technological sublime. Images and discourse focus on the sheer size and power of the accelerator, the complexity of the detector, and the computing power required to process vast amounts of data work in concert to point viewers back to the sublime purpose of the laboratory and its technologies.

But what of visitors' responses to such rhetoric? How do nonscientists make sense of the exhibits, models, and films that make up the self-tour? If the laboratory's rhetoric cannot engage its publics—if the texts, images, performances, and experiences fail to generate public support for the lab or stop challenges to the authority and autonomy of the high-energy physics community—new attempts to stabilize the boundary between the laboratory and its publics must be made. Thus even as the laboratory attempts to respond to public attitudes and beliefs about basic physics research, it is never possible to move beyond the process itself. The laboratory, its employees, and its publics are constituted inside an unfolding rhetoric that forms the very boundaries and spaces in which they exist. The remainder of the chapter is therefore devoted to nonscientists' perspectives on the self-tour, the laboratory, and the larger contours of the science/society relationship.

Interpreting the Laboratory

To assess public attitudes and beliefs, this chapter draws on interviews conducted with twenty-seven visitors to Fermi National Accelerator Laboratory as well as observation and analysis of community relations across the national laboratory system.[18] Interviews were also conducted with three visitors to the last FNAL open house, held 15 September 1997.[19] Because the presentation of the laboratory differed substantially during the open house, these interviews and five additional years of observational research have been treated as supplemental information that provide additional context for the analysis of self-tour visitors. As a result this chapter forwards general observations and arguments about the experiences of laboratory visitors, but it also seeks to illuminate more discreet ways in which nonscientists find meaning in science and therefore help shape the boundaries of science and scientific institutions.

The interviews discussed here map visitors' responses to their self-guided tour of Fermilab and also explore their understanding of the science/society relationship. Three dimensions emerge from this analysis of visitors' experience of science. First, visitors interpret the science/society relationship in terms of factors that have or may yet have an effect on development of research and its technological outcomes. Second, visitors distinguish between the expertise and power relationships that define the practice of science and the funding of scientific research. Third, visitors seem to experience the technological sublime differently than members of the high-energy physics community.

Challenging the "Endless Frontier"

Previous analysis of boundary work often gives readers the inaccurate impression that only scientists worry about or are threatened by changing boundaries between science and society. Contemporary debates about genetic engineering or missile-defense systems easily demonstrate, however, the degree to which nonscientists actively construct boundaries between themselves and science. Public resistance or controversy about developments in science and technology is, as the previous chapter demonstrates, often misinterpreted and misrepresented as a lack of public understanding, and nonscientists' opinions are too often classified as "correct" or "incorrect." Visits to laboratory visitors' center, self-tours, or open houses often open new spaces for dialogue, however. Designed to increase understanding, these encounters often prompt active contemplation of the boundaries between scientific institutions and their publics as visitors contemplate social changes wrought by the development of science and technology over the years. This mode of interpretation is rarely the intended effect but is instead the product of the visitors' presence within the laboratory space and their direct engagement of its discourses. Thus visitors may engage carefully designed texts and displays, but their interpretation is driven by stocks of knowledge that may or may not overlap with the interests of the laboratory community. The following exchange is indicative of this process. The self-tour at FNAL makes no mention of cold-war issues, and yet many visitors necessarily linked federally funded physics research to its origins in the Manhattan Project and the nuclear-weapons complex. Consider how three participants debate the respective responsibilities of science and society and, in the context of the laboratory, reify the link between physics research and weapons of mass destruction:

> V20: "Some of it's real scary. I mean, we were all scared when the atom bomb was unleashed because we were afraid that in 20 years we were going to annihilate ourselves in WW3. . . . Came darn close."
> V18: "We've got big problems with that technology [and] the fallen Russian States."
> V19: "And the ones that have had it [the knowledge] for years but have they really produced a bomb or not?"
> V18: "That's not science, that's society."
> V20: "But we've blamed scientists because they were the ones who unleashed this knowledge on us."
> V18: "See, that's where society has to control itself, though."[20]

From the perspective of those in the laboratory, such connections are a misfire and a chance to correct a misinterpretation. Despite efforts to erase cold-war connections in its public displays, the relationship between the laboratory and the military-industrial complex was reconstituted within this and a number of other visitor conversations. These visitors called forth a narrative that emphasizes the boundaries between science and society, and science and technology, and calls into question the benign character of scientific knowledge. The slippage among terms is not the result of ignorance or a "category mistake" as some might assert, but represents instead a quite deliberate attempt to push back at the boundaries of the laboratory by reopening the debate about the role of physics in the world, particularly the historical relationship between the national laboratory system and the development of weapons of mass destruction.

Thus this example is instructive for a number of reasons. First, it illustrates the inclination of nonscientists to contemplate the science/society relationship. Secondly, it demonstrates how the lab's publics may comprehend knowledge production with a sophistication not generally attributed to nonscientists. Third, it demonstrates how visitors differentiated between the knowledge produced through scientific research and the responsibility that accompanies making use of this knowledge. Fourth, this small exchange encapsulates a range of possible interpretations of the boundary between science and society. Although the exhibits in the self-tour define the institution in terms of the science it produces, many visitors reversed this assumption, reading the science produced in terms of the system of institutions from which it emerges. Participants in the above quoted discussion also contemplated whether the atomic bomb is a social or scientific problem, a scientific or a technological problem, or a matter of the knowledge produced versus the application of that knowledge. Finally, this conversation implies a need to exercise control over science in order to protect society from unexpected and undesirable outcomes of scientific research.

Most important, however, are the boundaries themselves. Notice how the discussion, despite its critical tone, only underscores the science-society divide. Science and society are separate spheres. Society is afforded both power and responsibility in the responses, but the experience of science is characterized as a struggle with the "other." Science is neither fully responsible nor completely absolved of responsibility for the threat posed by the knowledge it has "unleashed" on society.

The violent imagery of threat and struggle that peppers the participants' discourse is significant and reappears in a number of other visitor conversations. Another pair of participants communicated similar concerns about limiting the influence of science in society. Their conversation is more abstract and starts simply enough with utopic descriptions of science as key to "our way of life." Note, however, how the conversation turns and the counternarrative of "dangerous knowledge" emerges:

> V26: "I think it [science] helps us develop products and develop our way of living . . . science is helping everything evolve, like from simple tools to more complex tools and tools that are more efficient."
>
> V27: "I don't know if science does all that . . . but I sometimes wonder if it makes our life too complicated. Sometimes our life gets too complicated."
>
> V26: "It concerns me too . . . trying to explain away some of the things that I believe we need to accept on faith. [Like] trying to discover how . . . the universe was created. My belief system is that God created it and I think it tries to explain away God."
>
> V27: "That's what I was feeling when I was looking at that movie in there . . . maybe we're trying to dig into the particles [that are] only to live to be so old anyway."[21]

These participants seemed more concerned about unintentional or unanticipated intrusions of science and technology into their established (in this case, religious) systems of belief. As a result their comments do not reflect a specific social controversy but echo instead what they experienced as an uncomfortable mixing of the public and the private in and through science.

They accepted the role of science in the development of technology, but they seemed concerned that science might replace belief systems they felt were better equipped to answer fundamental questions of origin. Their negotiation of the boundary between science and religion demonstrates a desire, similar to that of the previously quoted participants, to differentiate between scientific and social concerns. Curious distinctions emerge, however, between the responses of these and the previous group of visitors. The first conversation challenged accepted narratives of technology by focusing on the unintended consequences of technological development. The second group accepted the many benefits of science—defined in technological terms—but worried about the philosophical or spiritual costs that might result from certain forms of scientific inquiry. Both examples reflect a general tendency on the part of visitors to vest science and society with different responsibilities and thus underscore the boundary between the two. For these participants science generates products, both technological and intellectual, and society determines the application and interpretation of those products.[22]

This tendency toward a functional definition of science and toward a distinction between the production of knowledge and technology and the social application of that knowledge points to the first major finding of this analysis—visitors define the science/society relationship in terms of application of knowledge and the production of technology. The initial indications of this can be found in visitors' general appreciation for the role of science and its products in their everyday lives. Several participants characterized science and technology as tools. These participants, like many others, collapsed the categories of science and technology and offered several concrete, technological examples of the benefits or uses for science:

> V11: "[Science plays a role] in measurements and anything you do. There has to be a certain chemical balance . . . to make the cake rise. You get in the car and that's science, isn't it. It goes forward."
>
> V12: "Welding, concrete . . . dealing with the elements to make the concrete right."
>
> V11: "Communication . . . between countries."
>
> V12: "Medication."
>
> V11: "Well, I flew out here from Denver. . . . I mean, how could I have done that a hundred years ago? I couldn't have been here in 3 hours, so that's science."[23]

This couple, one of whom was actually involved in construction work on the main injector, provided an expansive yet "technologized" definition of science, one that would seem favorable to any number of research pursuits provided that work was strongly tied to the development of practical, technological outcomes. When asked their impressions of work at FNAL, one visitor (V11) questioned the application of the knowledge or technology developed through high-energy research and commented: "I would like to know what the benefits are of doing all this. . . . I really didn't get that from here [the self-tour]. Are they trying to do this for medical reasons? Are they doing it just to find out? What is it? What are they going to do with the information when they do get it?"[24]

This comment was indicative of a pattern that emerged in several interviews. Many participants actively sought tangible, often technological, outcomes from particle-physics research but had difficulty locating them in FNAL's rhetoric. Responses of this sort suggest how the mundane, in this case the technological, may function as a counter-rhetoric to the sublime. These visitors and others did not necessarily deny the sublime dimensions of Fermilab's work; they simply demanded something more.

Though many participants defined science and the science/society relationship in technological terms, others focused on the application of physics knowledge in other spheres of everyday life. One woman (V6) described how she used concepts learned in college physics to help her manage her household: "I have a background in chemistry and physics and I tend to relate these sciences to running the household. The way the heat system works and the way the electrical system in the house works . . . the rain running off the roof and how it's caught by the gutters . . . Unless you know a bit about this, you are not really capable of understanding even the workmen when they come to repair."[25]

This participant clearly felt she had derived a practical benefit from a body of scientific knowledge. The fact that she had taken science classes in college and felt a certain degree of mastery over scientific concepts gave her a sense of agency that extended to her experience of the self-tour. When asked for her impressions of the tour, she distinguished herself from the "average tourist" on the basis of her education and knowledge of science, saying: "I thought it was well presented. It moved a little too quickly for someone who has no physics background at all. An overview might help as an introduction. . . . I think I come to this with a little more background than your average tourist, having had at least college physics."[26]

Even as her mastery of basic scientific concepts seemed to empower her, it did not guarantee a positive or sympathetic assessment of the laboratory and its work. Instead, in contrast to assumptions about the "interested public" held by many of the scientists interviewed for this study, "interested" and "educated" visitors were not unquestioningly supportive of the laboratory. This participant and others often leveraged their knowledge and interest in science to push back, with varying degrees of intensity, against the claims and authority of science. This participant's scientific education seemed to make her feel more confident in her critique of the lab's role in society. Although clearly appreciative of the sophistication of scientific research being conducted at FNAL, she described the lab in the following terms "For a physicist, it's got to be the most fabulous playground. . . . Someone who really enjoys physics and the workings of matter and energy. It's got to be like having the most sophisticated toys that they can possibly have . . . and being paid to do the thing they love to do the most."[27]

Even though principles of physics had a practical application in her daily life, the laboratory seemed to her a playground, an extravagance better suited to the desires of physicists than the needs of society. The work of the laboratory was redefined as "play," a pursuit designed to satisfy an individual's desire rather than society's needs. As a remedy she proposed that the self-tour stress the application of the knowledge produced at FNAL and offered the following suggestion: "If I were to give any advice, it would be to point out a little bit more what all this is for; it's not just a bunch of physicists having fun in an exotic playground . . . that there's a purpose. This is our tax dollars and it would be nice for people to be able to understand how maybe their lives would be improved by some of the research done here."[28]

Other visitors shared these concerns about the outcomes of particle-physics research and made similar suggestions. As one individual stated, "So you get down to the smallest particles . . . what are you going to do with them now that you've got them?"[29] This participant (V5) persisted in his questioning throughout the interview: "I am interested in finding out what value the whole thing is to me. It is a tremendous expense just to satisfy someone's curiosity . . . that's all I'm saying. I want to know what has been produced down here."[30]

In terms similar to those used by participant V6, this individual characterized the lab as something designed "just to satisfy someone's curiosity." Unable to locate an application or technological outcome amid arguments that emphasize sublime mysteries and the creation of knowledge for knowledge's sake, he assumed the personal intellectual curiosity of a small community of researchers motivated such a "tremendous expense." In frustration he explained: "I'm trying to understand what goes on, but I don't know. I've come here and have gone through it and I still don't know."[31] For this participant, understanding the presented material clearly entailed seeing a relationship between the knowledge produced and a specific social application or outcome.

In an effort to locate the source of his misunderstanding, this participant (V5) focused on the presentation of information in the self-tour and how it obscures understanding of any practical benefits that might result from the research. "Until I know how everything operates, I couldn't understand it . . . in laymen's terms, 'this is how the thing operates.' But they get into all these various, different names, even the name quark had me stumped for a while. You have all these different parts and particles. . . . They should all be explained. I used to build water filtration plants. The average person doesn't know how a water filtration plant [works] so I made a simple diagram . . . and I think it got through to a lot of people. If I had gone in there and given them a lot of technical stuff they wouldn't know what I was talking about. The same thing can be applied to this."[32]

As he critiqued the rhetoric of the self-tour, this participant also asserted his own rhetorical authority. He identified the source of confusion in abstract and poorly defined terminology and, like many other participants, suggested a remedy that drew on his own experience as an engineer.

Resistance and Redesign

This participant's response was similar to others in that it demonstrated the way in which many visitors actively sought out specific, practical explanations of and justifications for the research done at the laboratory. Many felt comfortable suggesting ways to improve the self-tour by making the displays easier to understand and the relationship between particle-physics research and society more obvious. As one participant commented, "Some of it gets deep pretty quick."[33] Another group of visitors elaborated on this concern, focusing on what they felt was excessive, poorly defined terminology:

> V20: "I want to see an exhibit that defines exactly all of these items. Just a basic definition of terms used throughout the entire exhibit here. Something I can look at [while touring]."
>
> V18: "Some of the poster presentations were too wordy [and] needed more illustrations. Just break it down into a few pictures and a few words about each picture."
>
> V19: "Warm us up!"[34]

Another couple, visiting with their children, reiterated the need for clarity. They suggested the tour should increase its emphasis on practical applications and decrease abstract references to theory:

> V16: "I think it was a bit too complicated for everyone. They should be more elementary first."
>
> V17: "They should put more practical things in here . . . because a normal person from a household can understand that better than all the $E = mc^2$ type of concepts. Neutron therapy, that's the kind of thing they should put more in the videos."[35]

To generate more interest and promote comprehension, another participant suggested a shift from reading to doing: "More interactiveness as opposed to reading . . . more than just reading a bunch of plaques on the wall."[36]

On the surface, these comments reveal two principal concerns shared by many visitors: that descriptions and explanations be presented in terms understandable to the layperson and that socially useful outcomes be emphasized whenever possible. Considered more closely, they also reveal the degree to which visitors' interpretation of the self-tour involved mentally redesigning the exhibits. Rather than uncritically accepting the presentation placed before them, they took and remade the exhibits according to their values and interest in the laboratory. In doing so, they often transformed the laboratory from how it was presented in the exhibits into what they thought it should be.

The foregoing comments also provided clues to the ways that visitors push back at the boundaries of science as they interpret FNAL's rhetoric. The fact that the visitors seemed willing to suggest changes to the exhibits speaks to specific aspects of their interpretive experience. First, visitors actively interrogated the texts they encountered. They seemed to regard the tour as an opportunity to both educate themselves about the lab and to place FNAL within the larger context of publicly funded research facilities. Second, most visitors interviewed came to the experience with a narrative of science already in mind—frequently a story that emphasized science as a source of intellectual or technological commodities. As they toured FNAL, they interpreted the work of high-energy physics through this lens and searched for tangible outcomes or products that result from the research undertaken at the laboratory. The self-tour, however, provided few resources for this purpose and was thus a frustrating experience for many visitors. Failure to find personally relevant meaning in the self-tour often resulted in a retrenchment of the visitors' beliefs rather than a change in their understanding of science and the science/society relationship. Contrary to the beliefs held by many of the lab's employees, simply presenting more information about the lab and its work did not necessarily engender support. Visitors were not—and are not—blank slates, and when confronted with exhibits that did little to connect the work of the lab to their lived experience, they often pushed back against barriers to their involvement by discussing exhibits, asking questions, and making suggestions for ways to improve the tour.

Corporations, Democracy, and the Science/Society Relationship

Not surprisingly, visitors' definitions of science and the science/society relationship shaped their perceptions of the function of the laboratory in contemporary society. Visitors' comments reveal a subtle but important distinction between *doing science* and *setting science policy*. Although most visitors did not presume to offer an opinion on the day-to-day work of researchers at FNAL, they felt quite comfortable assessing the social worth of that knowledge produced by the lab. In doing so, many returned to earlier discussions about the payoffs for such research. One participant (V17) pointed out that basic research cannot guarantee the nature of its outcomes: "These kinds of things are very subjective because there is no quantitative thing you can say, 'Today they are doing this, tomorrow I'm going to benefit from it.' That really doesn't happen."[37]

Clearly this participant questioned the likelihood of realizing practical benefits from particle physics research. More important, this response interprets "research and development" in cause-effect terms that may articulate with the laboratory's assertion that basic research has an important impact on GNP while simultaneously refusing to say how this happens. The lab's rhetoric defines the social impact of high-energy physics through two somewhat contradictory argumentative threads—economic impact and cultural value. Visitors seemed confused by the conflicting tales. On one hand the exhibits seemed to claim that physics had a definable impact on the U.S. economy and that this was a strong reason to support its work. On the other hand visitors were bombarded with messages claiming that science should be supported for its cultural value alone, as an expression of our humanity regardless of its practical applications or impact. In response to conflicting messages, many visitors rejected one or sometimes *both* claims. Drawing on the economic impact narrative, one visitor (V1) openly questioned society's investment in basic research by talking in terms of "rate of return": "This is a very low rate of return. All of the amount of money used and the next step so expensive.... Were I asked (my opinion) the answer would be that the practical application of such an expense is very limited in the near term."[38]

The "knowledge for knowledge's sake" cultural value story was similarly rejected. Many visitors found it difficult to regard scientific knowledge as sufficient justification for the vast expense associated with high-energy physics. Furthermore some questioned whether knowledge was always an undeniable good, occasionally describing the research in somewhat Faustian terms. V7 stated, "I don't see finding out about quarks and everything else as critical to [society] or a particularly [important] contribution. I suppose it's just a human need to understand all that, [but] would we be worse off if we didn't understand? Probably not. And are we meant to understand? I'm not sure about that either."[39]

Though many of these same participants felt the self-tour to be positive experience, describing it as "impressive," "interesting," and "very well presented," this positive regard did not insulate the lab from criticism of its perceived lack of social relevance nor did it reassure visitors that scientific knowledge was a neutral commodity to be produced and consumed without consequence. Most visitors expressed high regard for scientists and basic research, but admiration did not always imply support. The preceding comments reveal an audience that was actively searching for, but having difficulty locating, what it could recognize as good reasons to support the laboratory. Most participants were cognizant of the common arguments in favor of basic research, but they desired more explicit demonstrations of social benefit. On the whole participants' comments seem to indicate that nonscientists were more discriminating and more willing to contemplate the consequences of research and weigh the relative value of various projects than previous research in the public understanding of science might seem to suggest.

Their efforts to interpret the presented information and carefully weigh their regard for science and scientists against the relevance and technological promise of research contradicts some scientists' characterizations of the public as gullible and largely devoid of critical thinking skills. All participants expressed the desire to learn more about the lab and the social, scientific, and technological issues that affect its funding. Contrary to physicists' expectations, the strongest critiques of particle physics were offered by those with the most education, experience, or previous exposure to science. Those with less experience or knowledge were often supportive of research, but if they criticized the lab they were careful to qualify their conclusions in the absence of more information. Thus the judgments of these visitors, while forthright and sometimes unfavorable, were neither emotional nor *irrational*.

The visitors' tendency to distinguish between favorable regard for science and financial support for scientific research has important consequences for the understanding of boundary work. Most theorists contend that boundaries exist to reinforce separation between communities. However, when communities are constructed in mutually dependent terms, the rights and responsibilities of each group can and do come into conflict. In these cases critical distinctions must be made as to whether the allocation of research funds is a right of the scientific establishment, the tax-paying public, the federal government, or some combination of all.

These distinctions can be better understood by examining the implicit models of decision making that frame the discourse of scientists and nonscientists alike. The science/society boundary is managed through a social contract that negotiates power relationship in terms of decision making. Most physicists interviewed argued that decision makers should be required to demonstrate knowledge and familiarity with scientific method as a condition for participation in public deliberation about science policy. While such qualification could in theory be met by a nonscientist, FNAL rarely referenced public participation in policy making. In contrast many visitors gave priority to one's status as a member of a participatory democracy and flattened the hierarchy that some physicists sought to preserve. While visitors admired those who worked at the lab and distinguished them from nonscientists on the basis of their talent and intellect, many denied scientists significant authority in matters of decision making. Scientists and citizens were viewed as equal partners in the decision-making process, each with a different role to play.

Discussion about public participation was influenced in part by stereotypes held by both those inside and outside the laboratory culture. As they tried to imagine what a truly public dialogue about science policy might look like, they necessarily imagined "the other" with whom they might interact. When asked to imagine the sort of people who work at the laboratory or to describe a typical day inside the lab, visitors often focused on difference. The following exchange typifies the responses given by several visitors when asked to describe the people and the working environment at FNAL:

> V26: "It's probably a lot of tech-minds, tech-heads."
> V27: "They're into their area of expertise more than personal relationships. . . . We might be surprised how nice they are."
> V26 (joking): "We saw one guy in the video and said, 'He should get out more. He should go to Batavia.'"[40]

This exchange differentiates between scientists and nonscientists, invoking stereotypes that seem to challenge scientific authority. These participants marked the differences between scientists and nonscientists by speculating about the scientist's single-minded interest in research. As "tech heads," the scientists, with their capacity for balanced judgment, revered in scientists' rhetoric about democratic discourse, are called into question. Visitors seemed to admire and identify with what they perceived to be scientists' passion for their work but questioned their objectivity when it comes to science policy. Visitors also challenged the hierarchical relationship between scientists and nonscientists by suggesting that working at FNAL might be "just another job," not unlike their own.

This alternative vision of democratic discourse was, not surprisingly, colored by contemporary rhetorics of the marketplace. Visitors often positioned themselves as "citizen-consumers" of science. When asked to describe the working environment at Fermilab, one visitor (V22) commented that a typical day would probably be "coming in and analyzing information and solving problems . . . probably similar to other fields, it's just different in nature . . . just different topics. I would compare it to a large organization like a corporation."[41] Several participants compared FNAL to a corporation, transforming the priestly rhetoric of the institution into something more closely resembling a Dilbert cartoon. Another pair of participants elaborated on the corporate metaphor noting the different types of workers required to make the facility function efficiently:

> V18: "I think it's kind of like a big company atmosphere."
>
> V19: "That's what I was going to comment on. . . . There is the director and immediately below the director what I would call 'the thinkers.' But there's an awful lot of administration . . . somebody's got to keep the elevators running. It's like any other company that I've ever worked with."[42]

The tendency to transform FNAL into a corporate space provides further evidence of visitors' pragmatic resistance to the laboratory's attempts to justify the value of its research in sublime terms. Rather than discuss the search for fundamental particles in terms of the sublime, as the lab's rhetoric would suggest, visitors likened the atmosphere to a large corporation, one where many—not just a select few—labor toward a specific and tangible outcome defined by the marketplace of scientific knowledge. Given this image of scientific work, it is not surprising that visitors conceived of a role for themselves in decision making about science policy and the future of the laboratory. It is important to note that these visitors did not presume to make such decisions single-handedly, but rather with the help of working scientists. As citizen-consumers they remained dependent on skilled producers to delimit a range of possible product choices. Thus when confronted with a text that worked to deflect rather than address the questions and demands of a consuming public, many visitors were frustrated in attempting to enact their roles in the decision-making process. For many visitors, "democratic" participation was defined in part by the opportunity to evaluate the social worth of the laboratory's research and decide whether or not to "buy in." Having been given no choices to contemplate or alternatives to compare, the experience of the self-tour often engendered *resistance* rather than building support. Despite some physicists comments regarding "taxpayer choice," the self-tour experience did not produce any meaningful sense of agency for visitors. Although most came seeking both knowledge and an opportunity for engagement and input, many left feeling that they were allowed to be at best a passive witness to the lab's work.

Visitors' interpretations of the self-tour and subsequent critiques of the laboratory reveal the degree to which those traditionally thought to be outside of the laboratory have power to shape its boundaries. These participants' characterizations of the laboratory harken back to the earliest debates about the science/society relationship, but they also anticipate recent developments, including the increasing trend toward corporate management of the national laboratories. Having entered into what they viewed as dialogue with a scientific institution, visitors fought to participate, even in the face of texts that seemed to encourage silence and deference to scientific authority.

Domesticating the Sublime

As much of its discourse reinvigorates the sublime, Fermilab seems to reach back to a time when the boundaries between science and the public, between science and the federal government, and between laboratories and universities were clearly articulated and control over science policy and funding was fully vested with scientists. The powerful discourse of "consumer choice" seems to threaten the rhetorical potential of the lab's strategy. While visitors to FNAL expressed genuine fascination with the complexities of high-energy physics research, recurrent use of "corporate" and "consumer" metaphors indicated the dominance of this rhetorical frame and suggested that many visitors regarded FNAL as just one enterprise among many competing for taxpayer dollars. Thus the rhetoric of the marketplace trumps sublime experience and claws at the hierarchies it implies. As it introduces a new set of metaphors, it undermines the once-priestly authority of laboratory scientists and vests power in laboratory "stakeholders" instead. Most of those interviewed admired and had high regard for the work being done at the lab but did not necessarily accord any special status to high-energy physics in comparison to other scientific projects. In fact many talked openly about the need to support health and environmental research before allocating funds to the basic sciences. Though they read with interest those explanations and justifications made available during the self-tour, they pushed back at the historically established boundaries that mark the territory of high-energy physics by demanding pragmatic and technological justifications for research being done.

The mismatch between rhetoric and audience becomes obvious in competing interpretations of the sublime. Physicists described a special sense of awareness and control that grows from an understanding of the mathematical balance of the subatomic world. Both their rhetorical and technological creations are designed to reflect and to augment this level of understanding. Thus physicists and lab employees described a deep-seated belief in their work and a sense that the sublime infiltrated high-energy research on many levels. Visitors, in contrast to lab employees, were not persuaded by the sublime. Visitors' interpretations seemed to deny the symbolic unity of theory and experimental technology; instead they viewed technology as a practical outgrowth of the research process. In short they were more likely to be impressed by the spectacle of the grounds, the architecture, and the lab's technology than the nature of the research itself. This finding is particularly significant among participants who self-identify as interested in science.

The lab's rhetoric was successful to the extent that visitors consistently commented on the benefits of having a natural area available for their use. Participants clearly recognized the challenging beauty of the site and appreciated how it differed from the surrounding suburbs. One visitor commented, "When we walked in, we thought it was so beautiful with the garden. . . . it's very inviting. It wasn't a sterile laboratory environment."[43] Another participant also commented on the beauty of the lab's environment, "It's more like a park . . . the geese and the wildlife. My husband told me about how beautiful it was but you can't picture that."[44] Comments such as this were common. Though reflecting appreciation for the laboratory, they also reveal that visitors experience the grounds in terms of "added value" rather than awe. Visitors' appreciation of the natural areas did not seem to trigger sublime responses or affect their interest in or interpretation of the laboratory's work. As one participant put it, "It's just such a nice natural area. . . . We're not really here to learn about quarks today, but it's been a pretty nice afternoon of visiting the grounds."[45]

Visitors seemed to be able to "domesticate" the natural sublime as it was expressed through the lab's grounds, but the technological sublime was far more problematic. Sensing the apparent lack of practical outcomes from particle-physics research and lacking the experiences that motivated physicists' interpretations of their work, visitors could not integrate the natural and the technological sublime. This rupture in interpretation left in its wake a discursive space in which visitors felt empowered to challenge the claims and authority of the scientific community. It should be noted that the sublime rhetorics did have a limited appeal for some of those interviewed. As the comments presented in this chapter suggest, however, even the most ardent "physics fan" was capable of producing multiple interpretations of the laboratory—interpretations that reflected his or her complex positioning in the larger culture.

When compared to laboratory employees' comments, it became clear that the boundaries of the lab were and are destabilized by competing interpretations of democratic practice. As earlier chapters detail, this tension dates back to the earliest debates about the national laboratory system and is also reflected anew in the DOE's most recent management and oversight initiatives. Unfortunately the present situation has produced a stalemate. Attempts to promote "open dialogue" and "public communication" across the system seem only to reinscribe the boundaries between science and society. Ironically this problematic discourse may be the only common ground to be had in the debate. If they so choose, the various parties involved can choose to reject both models—elite governance and consumer choice—as equally unproductive and engage new modes of communication. The final chapter examines how we might begin this process of change through engaged research and cooperative action in order to blur productively the boundaries that currently, if the evidence presented in this chapter is sound, divide science and society.

⟫ SIX ⟪

Mapping the Boundaries and Charting a Future

The preceding chapters present historical, textual, and interview-based analyses that demonstrate what seems to be widening gap between Fermilab and the citizens whose tax dollars fund its operations. Although both groups seem genuinely motivated to improve communication across the divide, traditional approaches—monologic forms of education, explanation, demonstration—only reify the boundary that defines the ideological differences between science and society. Those inside the laboratory valorize its work as both scientifically and culturally significant regardless of cost or outcome. Those outside the laboratory demand some sort of tangible return on the investment made with their tax dollars. This ideological divide gives way to mistrust, misunderstanding, and struggles for identity and control. Physicists and laboratory insiders dismiss public concerns, critiques, and questions as uninformed or based on a fundamental misunderstanding of the nature and process of scientific inquiry. Visitors are equally perplexed by and resistant to scientists' insistence that the lab's research should be seen as a cultural and governmental imperative. Each group seeks input from the other even as they hope to exert more control over the funding process.

What is lost in this struggle is our understanding of the contingent nature of the boundary itself. The divisions experienced by each group are a product of long-standing cultural narratives that have created the science/society divide and as a result have perpetuated decades of often-unproductive debate about the role to be played by science and scientific institutions in the larger culture. Those inside and outside the laboratory's boundaries act in ways that reflect their narrative stocks of knowledge. Insiders dream big, imagine the unimaginable, and thus fulfill the expectations of their professional culture and those of the larger culture. Outsiders function as watchdogs, ever mindful of the comparative costs and benefits that result from favoring one type of research or one mode of cultural experience over another. As a result the sublime rhetorics forwarded by the laboratory are often challenged by skeptical stakeholders. With each exchange the boundary around the laboratory is reinforced, and the possibility of finding new ways to communicate about Fermilab's future is diminished. In this final chapter, I hope to summarize some of the most important observations from the preceding chapters and contemplate ways to move beyond this communicative impasse.

Ties That Bind: History, Sublime Rhetoric, and Everyday Life

Historical analysis provides valuable insight into the interactive, relational nature of FNAL's rhetoric. As previous chapters detail, high-energy physics has had a dynamic and sometimes-turbulent history. Over the years the field has struggled to be recognized as independent from the military-industrial complex from which it emerged. The energetic crafting of this distinction has always been tempered by the knowledge that, divorced from its military-industrial roots, the field might lose the very social relevance that once fueled its growth. To address this risk, the physics community has crafted alternative justifications for its work, justifications that embody what physicists believed to be the most effective model for the control and oversight of government-funded basic science.

Following Vannevar Bush's "best science" model, the physics community constructed rhetoric that initially emphasized potential economic benefits as well as the cultural significance and power of science. The resulting rhetoric linked scientific achievement to human achievement by comparing scientific and technological creations to the great monuments of Western culture. In contrast to competing rhetorics that stressed the importance of applied research and democratic governance of science, early narratives of the "new frontier" valorized the distinctiveness of the scientific ethos and stressed that the direction and practice of research must be controlled by scientists to be effective. Over the years debates about the role of science and publicly funded scientific research continued, and strong distinctions were repeatedly made between so-called elite and democratic models of governance and decision making. An uneasy stalemate eventually emerged in which powerful members of the scientific community retained a high degree of social prestige and control over the direction of research while research budgets gradually shrank in real dollar terms. Facilities such as SLAC and Fermilab were able to survive through a series of well-executed research programs and carefully planned upgrades. But the next machine, the SSC, did not fare so well as debate over the future of physics research roiled among government representatives, popular media, and members of a divided physics community. When the smoke had cleared, it seemed that debate had moved beyond competing models of governance to reveal a widening gap between science and society. Sensing this, high-energy physicists have contemplated ways to make physics more appealing to both the general public and those controlling government funding.

In many ways Fermilab has been at the forefront of efforts to revitalize high-energy physics rhetoric. The laboratory first developed its unique aesthetic rhetoric as a response to the funding crises of the late 1960s. Robert Wilson's effort to fuse traditional "best science" values with a mythic rhetoric of the sublime worked to distinguish the lab from the military-industrial complex in an effort to build a secure foundation for both Fermilab and future research facilities. In the late 1990s, while considering its future in the wake of the SSC cancellation, the laboratory community looked to its rhetorical roots to rebuild public support by attempting to make physics seem once again interesting, exciting, and culturally significant.

The resulting rhetoric was largely visual and experiential in nature and, as it had in the 1960s, seemed to stand in contrast to the public's stereotypical understanding of physics research. Emphasis on art, architecture, photographs, and constructed views of experimental technology offset perceptions of the abstract and mathematical nature of the research. Fermilab's construction of the sublime emphasized a dynamic tension between beauty and power—the complex beauty and power of the accelerator and detector technologies and also the beauty and power of "natural" subatomic particle interactions. Of these two the technological sublime was dominant, but the combined influence of the natural and the technological seemed to serve two needs. First, it held the promise of improving high-energy physics' position with respect to its perceived competitors in the funding race by demonstrating that physics was capable of generating profound beauty and power. Second, the approach allowed the lab to focus on its technology and yet simultaneously deflect, when necessary, discussion of the practical applications of its work.

The analysis presented in this book provides evidence of a historical pattern of response that seems to characterize not only much of FNAL's boundary rhetoric but, more generally, public communication of science. The course of physics' history demonstrates the recurrent reinscription of the science/society divide, particularly when institutionalized science is challenged through controversy or other forms of public pressure. From the time of Harley Kilgore through to the present day, the high-energy physics community's valorization of pure science has consistently been met by calls for consideration of practicality and the public interest. Even the sublime appeals of the space program—historically some of the most dramatic and quasi-religious scientific rhetorics—have given way to more pragmatic appeals under the weight of immediate public concerns about efficiency, practicality, and productivity.

In the case of Fermilab, the rhetorical pattern is well established. When budgets tightened or public attention was fixed on other concerns, calls went out to justify the expense of particle physics research. What this demonstrates is not so much the failure of lofty appeals to the beauty and power of particle-physics research, but rather the physics community's reflexive return to a sublime rhetoric when attempts to address pragmatic concerns have failed. The push and pull of competing interpretations of the social role of scientific research gives structure to the boundaries between science and its publics. Decision making once reserved for the power elites of the physics community is now taking place in the halls of government, and public opinion is becoming increasingly important in these deliberations. A public that once stood by and watched as national laboratories were created is now considered a stakeholder in those laboratories. Over time roles and relationships shift and members of the physics community must adjust their understanding of the work they do based on the perceived level of public support for high-energy physics. Thus the public response to Fermilab's rhetoric is of critical importance as the community attempts to anticipate or preempt further boundary incursions that appear to threaten its future. The permeability of the line separating Fermilab from its publics and from competing research institutions is never entirely fixed but is in a constant state of flux as it responds to developments of both sides.

The patterned relationship between Fermilab and its publics illustrates a shortcoming in previous scholarship employing the "boundary" metaphor. While the separation of science from other social institutions is a central component of the rhetorical construction of its authority, the placement of the border is not the only issue to be contemplated. The *permeability* of the boundary is equally important. The degree to which "boundary crossings" are allowed and the conditions under which such incursions occur tell us much more about the nature and health of the science/society relationship than do simplistic observations that the boundary exists.

Since it is only in its earliest stages, it is difficult to gauge the effects of Fermilab's latest invocation of the sublime or its long-term usefulness as a strategy to promote high-energy physics. On one hand the increasing influence of the public and their apparent resistance to some aspects of the lab's sublime rhetoric could signal a breakthrough in the science/society relationship and movement toward a more productive balance of power. On the other hand this same resistance could simply imply a habituated response, the "take" in a give-and-take rhetoric of publicly funded scientific research. Funding for physics remains flat, but prospects for future funding will ebb and flow with changes in the White House and in Congress. At the present time, the majority of research funding still goes to the biosciences and other branches of research that have in the recent past adapted many of the rhetorical and scientific practices pioneered by physics to serve their research and funding goals. As visitors' comments reveal, concerns about the practicality and relevance of high-energy physics persist. While the sublime still infuses the majority of Fermilab's public texts, the lab's most recent efforts seem to feature a limited number of pragmatic arguments likely intended to assert physics' relevance in the increasingly competitive world of government-funded research.

It is somewhat more difficult to say to what extent sublime rhetoric permeates the whole of physics research. It seems certain that Fermilab's utopian rhetoric has exerted a strong influence on the course of high-energy physics and its perceived power as an institution. Comparisons across several national laboratories suggest that Fermilab's rhetoric is the culmination of system-wide efforts to differentiate high-energy research through appeals to aesthetic experience. Early evidence of this can be seen at predecessor laboratories, such as SLAC and Brookhaven, where art or cultural events were used to redefine the laboratory's relationship with the surrounding communities. Although several similar rhetorical strategies predate the founding of the Fermilab, the holistic design brought forth by Robert Wilson, Leon Lederman, and the Fermilab community was the most complete statement of the cultural significance of high-energy work. Looking inward to the core of beliefs that united the community and outward toward public fascination with science and technology, the Fermilab approach seemed intent on reasserting the authority of high-energy work in both the wider discipline and the larger culture.

The importance of experience—both rational and aesthetic—to boundary work is evident in FNAL's public texts and also in the responses generated through interviews with physicists, lab employees, and visitors. For physicists and lab employees, clear communicative connections exist among the many dimensions of lived experience—laboratory life, the culture of high-energy physics, the practice of research, and the production of the lab's rhetoric. For visitors the very act of interpreting the self-tour requires them to draw on an understanding of science that is grounded in their own lived experience.

Previous chapters reveal a professional community that justifies its existence in terms that are arguably more important to those inside physics than those outside. At a time when those inside the field feel that its very existence is threatened, the boundaries that are constructed need to reinforce the values of the community. Interview transcripts demonstrate how the day-to-day work of high-energy physics produces a rhetoric that is not simply a strategic choice, but also the product of a belief system that orients physicists toward their field.

Visitors described a very different experience of science than that described by physicists and lab employees. For most, science was the source of practical technology rather than metaphysical understanding. This characterization of science, born out of the experience of living in a technological culture, seems to marginalize physics among the sciences and accord greater importance to those fields that produce knowledge or products that can be consumed by the general public. While visitors recognized and often even accepted the differences between scientists and nonscientists brought forward through the rhetoric of the self-tour, their experience dictated that they judge the laboratory in terms of practicality and social relevance. Visitors' lived experience clearly influenced their interpretation of Fermilab's institutional ethos, and as a result most visitors did not hesitate to critique the lab and its work if they felt it did not speak to *their* understanding and experience of science.

In and of themselves, such differences in interpretation are not problematic given the polysemy of science in contemporary culture. The vast *distance* between the perspectives of those inside and outside the laboratory, however, is cause for concern. If public discourse about the future of science is to move forward, some common ground must be found between the sublime mythologies of the physics community and the increasingly powerful consumerist model that is evident in the comments of nonscientists.

Rethinking our historical patterns of public scientific communication challenges many problematic assumptions about the nature and seeming intractability of the science/society divide. For too long researchers have framed public participation in scientific decision making almost exclusively in terms of "public understanding," as though the only legitimate role for nonscientists was to understand and, by implication, to appreciate the work of scientists. While I have no desire to reject the importance of work in public understanding research and concomitant efforts to improve both formal and informal modes of science education, I do hope that the findings presented here will stretch our interpretation of the nature of public understanding and participation in science. The comments of various members of the Fermilab community reveal the degree to which many in the sciences hold stereotypical and often inaccurate opinions about the public's interest level and capacity for intelligent deliberation about scientific issues and the fate of scientific institutions. In contrast the visitors' comments demonstrate their willingness to both educate themselves about scientific issues and institutions *and* participate in a larger cultural conversation about the value and social significance of various kinds of publicly funded research. The very fact that so many individuals voluntarily tour FNAL on a yearly basis suggests that nonscientists of many stripes find the facility and its work sufficiently important to merit their attention and thoughtful consideration. Structured comparison of the experiences and rhetoric of these two groups, in conjunction with broad historical analysis of both the laboratory and the system it inhabits, suggests that the lab's decision-making processes are inevitably shaped by communication between the laboratory and a wide variety of real and potential audiences.

Those concerned about the past and future of organized research and scientific institutions can learn much through skillful observation of the rhetorical boundaries that define the nature and purpose of high-energy physics research. Engaging rather than denying this interactive process provides scientists and nonscientists alike an opportunity to reconsider the often-strained relationship between organized research and the publics that fund such work. Many of the comments included in this book demonstrate the problems associated with defining the science/society relationship in either dependent or adversarial terms. But even as they mark the divisions and the problems that result from these divisions, they suggest the possibility for alternatives.

A New Model of Interaction

The findings presented in this book both reinforce and challenge some commonly held assumptions about the rhetoric of science and the rhetorical construction of boundaries around social communities. In doing so, they also point to the need for and benefits to be derived from more research focused on public communication about science. The combined use of textual analysis and interpretive interview analysis presented here suggests new possibilities for exploring the rhetorical boundaries of contemporary science and the influence of discourse on our perceptions of the science/society divide. As research institutions such as Fermilab continue to evolve, increasingly significant questions about the role of science in society must be asked and answered. Critical analysis grounded in engaged research will be increasingly important if we hope to pursue such a dialogue and, in turn, better understand the social forces that drive the development of science in the twenty-first century. What is being suggested here is more than a simple call for more or improved research—it is a call for scholars and science communicators to reassess the nature and potential of their work.

First and foremost, scholars and science communicators must recognize that *rhetorical boundaries matter as much as physical boundaries,* primarily because they give structure and meaning to the relationships that sustain the laboratory. The divide between the laboratory and community is key to the constitution and continued existence of publicly funded scientific institutions committed to producing "valuable" research as measured in both scientific and social terms. Furthermore researchers and other interested parties must acknowledge that scientific institutions are *rhetorically constructed.* Regardless of one's opinion of the role of rhetoric in the practice of science, it is difficult to dispute that any given laboratory's identity and relevance in the larger culture is the product of either the collective or the competing rhetorics of scientists and nonscientists alike. Those interested in the future of scientific research need to focus their attention on the myriad ways in which research as social project and process is constituted through discourse. Denial or failure to attend to this dimension of the rhetoric of science will only exacerbate existing tensions that mark the divide between science and society.

The productive blurring of this divide can begin with the type of research presented here and in time continue through cooperative efforts undertaken by scholars and by those who work inside the laboratories. Far too often those charged with responsibility for communicating a given laboratory's mission to outsiders have little or no sense of their audiences. Some laboratories rely almost exclusively on large-scale public-understanding surveys for data regarding public attitudes and have little or no understanding of how the laboratory is viewed by its immediate neighbors. Few institutions have invested the resources or effort to conduct qualitative research that might provide richer and more consistent insights into meanings that nonscientists ascribe to science in general or to a particular laboratory and its research program. Scholars trained in rhetoric and communication, however, can help to forward a much more complex picture of the public. Narrative interviewing and rhetorical analysis holds much promise in this regard. Focus group and individual interviews framed by rhetorical analysis would provide new perspectives on how setting and social occasion affect the rhetorical construction of science. To the extent possible, interviewing in naturalistic settings (in addition to laboratory environments) would also provide valuable information. The information garnered through alternative approaches could then be combined with and compared to existing and new survey research in order to provide the broadest possible perspective on the nonscientists' relationship to the laboratory.

This book has, I hope, identified the negotiation of boundaries at a critical point in one lab's history, but longitudinal research is also necessary to provide an even clearer picture of the boundary-construction process. Future work in this area should track the changing nature, "location," and permeability of boundaries over long periods of time in order to build on a picture of the science/society relationship derived primarily from historical narrative. If we assume boundary work to be a truly dynamic process, then appropriately designed research can help to chart the significance and influence of historical events and interested publics over time and thus contribute to a more nuanced understanding of the nature and evolution of rhetorics of and about science. Such analysis could be expanded to include outsider discourses that contribute to the rhetorical constitution of the research and the community within the lab. Strategic analyses of local newspapers or professional journals would supply valuable insight into the emergence of specific types of claims and lines of argument that work to define the boundaries of the laboratory and structure relationships between institutions and their publics. In important historical cases, this type of analysis could be used to retrace the influence of outside publics and discourses on the boundary construction process of particular fields and institutions or on the evolution of significant controversies.

The foregoing discussion offers just a sampling of potential modes of research that might deepen our understanding of how and why we have come to live in a world defined by the divide between science and society. And yet the choice of methodologies is but one among many important considerations. The question of *who* is engaged in and by such work is of equal, if not greater, importance. As described earlier, exploration of public attitudes toward science has followed one of two models in recent decades. Public understanding and education research is framed by the traditional paradigms of social science in which the public is regarded as the subject of research. Despite some recent trends toward increasingly naturalistic and engaged approaches to such research, the subjectivity of the public remains largely unchanged. At the opposite end of the spectrum, engagement with the public has been mandated in the recent past in response to public controversy, most notably environmental contamination in and around national laboratory sites. Though not research in the formal sense, the creation of citizen boards or community advisory groups has produced a very different kind of knowledge of public attitudes toward science and scientific institutions than that generated through traditional social science. The promotion of public-participation programs by the DOE and the EPA holds the potential for meaningful engagement, and yet, because these encounters are rooted in controversies defined through opposing groups, the full potential for public dialogue is only rarely realized.

Between these two extremes, however, is a third possibility—an approach that draws on the spirit and intent of public participation to ask and answer questions about public communication and meaning. In such a model, the audiences to which the laboratories direct their communication are neither "target markets" nor students framed by the education model, but are instead partners in a cooperative exploration of the public meanings of science and its institutions. While scientists and nonscientists often do derive meaning from substantially different life experiences, there is space for cooperative effort in the meanings they construct for science. Future attempts to assess public understanding and attitudes must focus less on identifying the nature and sources of nonscientists' deficiencies in terms of scientific knowledge and devote greater efforts to removing long-standing barriers to nonscientists' participation in our larger conversation about the past and future of science.

The goal should not be to "fix" members of either group by demanding the "normalization" of the scientific community or the "scientization" of the public, but rather to create a new model for interaction in which the boundaries—if there are to be any—between science and society, laboratory and township, and science and nonscience are constituted in a spirit of cooperation and community rather than control.

Open Conclusion: Assessing New Challenges to This Model of Interaction after 9/11

This concluding section is an open invitation for new scholars to follow my footsteps in testing and elaborating the findings of this study. In a necessarily undeveloped and hasty way—which might not always avoid excessive but hopefully stimulating suggestions—I intend to indicate major challenges to the model of interaction that I have outlined so far. In particular 11 September and the changes in priorities that the war on terror generated reactivated the forgotten link between research in high-energy physics and its possible military applications and now represents a major issue newly confronting it. Directly facing these unprecedented problems will, I hope, contribute to the development of more open and mutually satisfactory relations between nuclear physicists and the general public.

Looking to the last decades of Fermilab, we could say that, in attempting to build spaces for pure science, those in high-energy physics never challenged the military foundations of the laboratory system. The sublime rhetoric they crafted was in the end inextricably bound to the terror of nuclear weapons. As Nye argues, in fact, the nuclear "sublime" is almost an abuse of the term as the atomic bomb does not allow for self-preservation in the feeling of Kantian sublime. Indeed it is the only technology that has never been successfully desublimated.[1] Despite its attempted domestification through the peaceful applications of atomic energy, the fear of the military applications of nuclear technologies never completely disappeared, doubling environmentalists' concerns about the safety of nuclear energy plants.[2] According to Nye, the thought of military applications of the nuclear sublime undermines the very possibility of a communal relationship to both natural and technological objects, hinders identification with them, and "transforms admiration for inventors, engineers, and scientists into fear and mistrust."[3] In the complex case of Fermilab this military legacy—undermining the careful construction of a sentimental sublime which I have discussed in the previous chapters—still seems to hold true, and has been reactivated in public consciousness by recent developments in the war on terror.

In sheer technological terms, the accelerator first received government support as a potential instrument in the country's war effort; the earliest basic research was indistinguishable from defense work. While few in the public know the detailed story of the accelerator's history, they are nonetheless conscious of the promise of technological spin-off, which permeates the past and present relationships between federal funders and researchers. In the public sphere, basic research is simply never quite basic research. Whether stated or not, the promise of payoff—often in the form of defense technologies—is always there. Hence even the most benign contributions of high-energy physics to computing and software are now being reshaped according to the needs of the war on terror. While I have no doubt that those in the lab are sincere in their desire to do only nonclassified, nondefense-related work, the rhetorical ghosts of the past have and will continue to haunt those involved.

Denial has been the dominant strategy employed to address this problem. At first glance, FNAL's rhetorical strategy seems innovative and creative as it seeks to redirect the question of basic research into the realm of aesthetics and the sublime. FNAL sought to deny its connection to the military-industrial complex through the creation of spaces and organizations supposedly unsullied by defense work—intellectual green spaces that would shelter both their occupants and their visitors from outside threats of contamination. Such a strategy seems to function effectively to resolve individual ethical and ideological angst. As a group FNAL employees seem convinced by their own discourse, secure in their denial and distance from the problematic past of the NLS. As a rhetoric directed at both critics and supporters alike, however, it is doomed to fail. As individual physicists reassure themselves that their work does not contribute *directly* to military-industrial interests, the very act of drawing the boundary only draws attention to the relationship. The lab that describes itself as *not* a weapons lab is locked in a binary with the lab that *is* a weapons lab. The lab that denies the possibility of environmental contamination is a constant reminder of the legacy of environmental damage that pervades the system.

The boundary—carefully and strongly drawn to deny any and all connections—can only highlight the linkages in the end. The knowledge and technology produced at FNAL can never be segregated from the world of weapons and defense in any meaningful way. Furthermore the same contaminants that have threatened other laboratories' neighbors are present at FNAL, even if they are not present in sufficient quantities to be hazardous at present. The recent controversy over a series of tritium spills documented between 1996 and 2005 at the Braidwood plants southwest of Chicago has raised public concern on this problem. Most important, the same funding stream that has flowed from government to the weapons labs over the course of history has also flowed through high-energy physics. In short there is no way to deny the historical, ongoing relationship between the worlds of applied-defense work and basic research. To do so is simply to deceive.

That said, I do not wish to argue that the labs are populated by a new generation of Dr. Strangeloves, whose ethical concerns have long since been overridden by maniacal devotion to science at any cost and who are interested only the preservation of their research agendas and technologies. I believe the scientists I met at FNAL and elsewhere to be sincere in their desire to avoid defense work and in many cases resist its influence on the funding process. Nonetheless they cannot see their way out of the Faustian bargain they have made, as the professional culture and practices of high-energy physics have become fundamentally reliant on machines that only the government can fund.

This dilemma leads some to suggest the equally unpalatable alternative of the marketplace. Private industry has shaped the trajectory of physics research since World War I. It can be argued that the government learned to pose questions of spin-off and technological development from the example of private industry. And thus the terms of the bargain are unchanged with private interests. Research can never stray too far from the question of its real worth in terms of goods produced.[4] Sadly, the comments from the public offered here seem to reflect these market logics. Just as the lab's scientists seem driven to deny the pragmatic possibilities of their work for fear of raising the nuclear specter, the public persists in questioning whether it is getting sufficient bang for its buck. And who can blame a public for ignoring legitimate arguments about the social and cultural value of basic research when those arguments are presented through a rhetoric of mystification? With boundaries drawn as they are and managed through the constant reinforcement of the rhetoric of the technological sublime, each side is inevitably drawn back to a foundational understanding of the bargain—funding comes only with the assurance of pragmatic (be they defense-related or not) outcomes. The discourse of the funding process makes this clear. The taxpayers are "stakeholders." Public money is an "investment." The devil will demand his due eventually. Furthermore each side operates from a position of mistrust. The public, expecting exaggeration or even deception, grows ever more skeptical of claims made to justify high-energy research and facilities. Those inside the labs resent and reject the public's involvement in the debate, questioning both their right and their competence to engage in the process.

In addition, if this lack of physicists' interest in the pragmatic applications of their work can be deemed a kind of apathy, in the case of environmental contamination the denial seems to assume a different form: they must constantly and actively deny the risk. To do this, they retreat into the realm of the ultrarational, refusing to admit any contravening evidence unless it meets a standard they would not likely apply in other realms of their work.

The greater the mystification—the stronger the boundary that separates the lab from the mundane world outside it gates—the deeper the divide. What began as an attempt to garner public interest and maintain autonomy in the face of government intervention has thus become a paralyzing habit. The maintenance of the technological sublime requires an increasing effort on the part of public-relations personnel and even a few lab scientists; what began as internally produced black-and-white brochures has evolved into full-color publications mailed to thousands every other month. Additional efforts had to be devoted to an elaborate revamp of the tour and increasing investment in community relations. Sadly, a strategy based in mystification encourages those in the physics community to deny the existence of (even legitimate) controversy and avoid public debate. As a result a community that claims to be investing heavily in its community relations is often surprised by the emergence of opposition or resistance to its activities.

Of equal concern the public often seems surprisingly complacent in its role as consumer of the laboratory and its products. A small number of citizens will routinely challenge the boundary in an effort to ask for more openness in matters that could affect the future of the community at large. These individuals, however, are seen as the minority, vocal or not, when compared to the larger percentage of the nontechnical public who have simply given up in the face the mystification. The logics of the marketplace have convinced "consumers" that passivity is a normative response as long as there is technological progress. A long-standing tradition of questioning this myth of progress seems to have been forgotten.

In the end, even in the changed configurations brought about by the war on terror, the goal of creating a "special space" need not be completely abandoned. A new space is needed—not one set apart from the needs and concerns of the mundane world, but instead one where those very topics can be addressed and engaged. A rhetorical scholar does not have to, and perhaps should not be, in the business of giving advice. The implicit lesson of this study suggests, however, that the scientists and the public interested in FNAL and in developing a more open relation of interaction should *give themselves* something like the following advice:

- The lab needs to abandon its dysfunctional, monologic rhetoric and stop trying to legitimate itself as "special." Greater opportunities for identification may result from the straightforward acknowledgment of the mundane nature of its work. What is mundane is often necessary. Owning up to the socially routine nature of scientific research would not necessarily diminish its importance in the scramble for public funds.
- The public needs to participate as well. We need a rational conversation about the promise and consequences of this type of research. Both scientists

and the public alike have become complacent, steeped as we are in the myth of American progress through technology. All too often, mystification denies the need for the debate *or* the logic of the marketplace reduces public "investment" to a matter economic scarcity. And yet, real questions remain unanswered:

> What are the possibilities and limitations of human knowledge?
>
> What are the real and potential consequences of this research and to what extent are we willing to accept those consequences?
>
> How have we addressed the consequences we currently live with and how might we learn from our mistakes?
>
> What is the relationship between our cultural values and science? Can we really agree about cultural values?
>
> How can we address social and cultural concerns alongside scientific concerns? How can all those involved honor (negotiate, discuss, etc.) their concerns through the decision making process?
>
> How can we widen the circle of involvement?

In the current rhetorical situation and climate of opinion, we can add to these general questions a number of additional remarks. The physics community is steeped, whether they care to admit it or not, in a sense of entitlement. This attitude derives from a long-standing contract with the government—an agreement that, although funding decisions are made in the name of the public, they are largely devoid of public involvement. The "public" is still a mostly fictional entity in the world of physics funding, and yet it is more frequently invoked. This is perhaps the greater proof that—in the newly changed national and global situation—the nature of the bargain is changing, making the construction of a productive space of dialogue even more pressing. Stakes are higher every year:

- Expenses grow as technological requirements grow.
- Recurring and expanding problems with environmental contamination still persist. The physics' community's denials of risk seems only to make the problem worse, to sharpen the divisions between lab and neighbors, and to paralyze what could and should be a meaningful and productive public discourse.
- The neighborhood in and around FNAL is evolving; property is becoming more valuable and the public increasingly resistant to anything that might threaten that tendency.
- Is it possible, I finally ask, to expose the boundary metaphor in order to envision a new configuration? Is the actual and metaphoric influence of globalization a factor in doing so? Is it possible that science and society will

no longer be separate countries on the map, but nodes in an evolving global system?

To conclude this work, I would like to indicate briefly some paths onto which future scholars might venture in order to deepen and enlarge my findings. In particular two large inquiries open up before us.

First, the possibility of a systematic project analyzing the comparative rhetorics of different national laboratories. We can hypothesize that though their rhetorical problems are often shared, their rhetorical situations, histories, stakeholders, missions, and institutional structures vary considerably and informatively.

A second worthwhile project would be to catalogue and chart in greater depth the effects of the security measures put into place after 9/11 on the rhetorical practices of national research laboratories. I would hypothesize that these are sure to have shifted the balance between publics and insiders on the one hand and between the pursuit of pure knowledge and socially necessary projects on the other. But I must leave these inquiries to future scholars.

Notes

Chapter One: Interactions

1. The national laboratory system encompasses a slowly growing number of laboratories and facilities. Having its origins in the Manhattan Project, the system includes widely recognized labs, particularly those associated with nuclear and physics research such as Los Alamos National Laboratory, Oak Ridge National Laboratory, Lawrence Berkeley National Laboratory, Lawrence Livermore National Laboratory, and Brookhaven National Laboratory. In the past forty years, two basic research laboratories have come to prominence within the system—Fermilab and Stanford Linear Accelerator Center. These facilities offer complementary methods for studying high-energy physics.

2. There is a tradition of magician cartoons at Stanford Linear Accelerator too, but these images are clearly intended to be humorous. The professionalism and solemnity of Gonzalez's illustrations points to the seriousness with which they were to be regarded. For many years Gonzalez's images provided the unspoken, mythic ground for the work of the laboratory. Her designs (most notably, the lab's logo) were seen on every brochure, lecture announcement, and official document. She documented every discovery or technological innovation with a new illustration. This way of treating the trope of magic reflects the rhetoric of the technological problems of Fermilab.

3. Chris Quigg, "A Little Bit of the Gods." "P-bars" refers to antiprotons.

4. Robert Wilson, *The Humanness of Physics*.

5. "History," http://www.fnal.gov./pub/about/whatis/mission.html (accessed 12 January 2006).

6. "Mission Statement," http://www.fnal.gov/pub/about/whatis/research.html (accessed 12 January 2006).

7. Ibid.

8. Ibid.

9. http://www.fnal.gov.pub/about/organization/budgetstatistics.html (accessed 12 January 2006).

10. While there have been few, if any, "rhetorical" histories of the national laboratory system, many fine narrative histories do exist. The following review summarizes

work by Daniel Kevles, Don Price, Peter Galison, Lillian Hoddeson, Catherine Westfall, and others that explores emerging relationships between the public and institutions devoted to basic research. The work of John Krige details the parallel evolution of high-energy physics in Europe. Such secondary sources not only provide valuable perspective on the factual history of various facilities and events, but also open a window onto the narrative life of these institutions. Many of the laboratories employ historians, either officially or unofficially, to document important events, maintain the laboratory's archive, and preserve, through both internal and scholarly publications, the stories of the institution. In the course of my interviews, I was struck by the degree to which participants attended to their institutional history. Sadly scientists' valorization of history was often counterbalanced by a profound lack of recognition of the funding and labor required to maintain archives and do historical research.

11. Lillian Hoddeson and Adrienne Kolb, "The New Frontier in the Chicago Suburbs," 2–18.

12. Vannevar Bush. 1945. *Science: The Endless Frontier*, NSF Web site, http://www.nsf.gov/od/lpa/nsf50/vbusg1945.htm (accessed 15 July 2003).

13. Peter Galison and Bruce Hevely, eds., *Big Science*, 14.

14. Mary Midgely, *Science as Salvation*, 3.

15. Leon Lederman, *The God Particle*, 225.

16. Daniel J. Kevles, *The Physicists*, 347.

17. Ibid., 347.

18. Ibid., 347.

19. Ibid., 345.

20. Ibid., 345.

21. Ibid., 348.

22. Ibid., 350.

23. Ibid., 350.

24. Don K. Price, *The Scientific Estate*, 3.

25. Kevles, *The Physicists*, 350.

26. Ibid., 351.

27. Ibid., 352.

28. Ibid., 356.

29. Ibid., 356.

30. Ibid., 361.

31. Ibid., 364.

32. Ibid., 365–366.

33. Ibid., 367; extensive descriptions of research programs at the UC Berkeley–managed labs can be found in the laboratories' annual report and also in copies of the labs' newsletter, the *Magnet*. The title of the newsletter provides ample evidence of the continued influence of E. O. Lawrence and accelerator-based physics.

34. Robert W Seidel, "The Origins of the Lawrence Berkeley Laboratory," 21–44. The received history of the Lawrence Berkeley National Laboratory emphasizes the "Rad Lab," and yet the early medical research conducted by John Lawrence with the

support of his brother, Ernest, was also clearly influential in the development of the laboratory. Close examination of primary documents from the period reveals the degree of attention given to medical research. The lab's contemporary research program has in fact shifted significantly in recent decades in the direction of medical, materials, and energy sciences. Basic physics research, particularly accelerator-based work, now constitutes only a small portion of the overall research agenda.

35. A history of the graphite reactor can be found on the Brookhaven Laboratory website, www.bnl.gov.
36. Kevles, *The Physicists*, 368.
37. Ibid., 369.
38. Stuart M. Leslie, *The Cold War*.
39. Ibid., 374.
40. Ibid., 374.
41. Daniel Greenberg, *The Politics of Pure Science*, 217.
42. Kevles, *The Physicists*, 386.
43. Ibid., 391.
44. Alvin Weinberg, "The Impact of Large-Scale Science," 161.
45. Price, *The Scientific Estate*, 11–12.
46. Ibid., 11.
47. Kevles, *The Physicists*, 393.
48. Price, *The Scientific Estate*, 12.
49. Kevles, *The Physicists*, 396.
50. Price, *The Scientific Estate*, 13.
51. Ibid., 14.
52. The NAL was renamed Fermi National Accelerator Laboratory (FNAL) in 1974.
53. Wilson, in testimony to the AEC Authorizing Legislation Committee on Atomic Energy Hearings, 17 and 18 April 1969.
54. I first encountered the phrase "kiss of death" during a presentation given in honor of Ned Goldwasser ("Reminiscences," presentation given at the Symposium Celebrating Ned Goldwasser's Eightieth Birthday, 10 March 1999.)
55. As with Fermilab, the Berkeley laboratory underwent several name changes. It began life during World War II as the "Radiation Laboratory" and was known during the postwar period as the "Berkeley National Laboratory." This designation eventually included two facilities, one adjacent to the Berkeley campus and one in Livermore, California. In the 1970s the two laboratories were officially separated. It was at this point that the Berkeley laboratory officially became "Lawrence Berkeley National Laboratory" and the Livermore facility was renamed "Lawrence Livermore National Laboratory." To avoid confusion, however, I refer only to the Lawrence Berkeley National Laboratory in this book.
56. "Mission Statement."
57. "Research at Fermilab," http://www.fnal.gov/pub/about/whatis/mission.html (accessed 12 January 2006).

58. As quoted in Mike Perricone, "Eye Witness to History," www.fnal.gov/pub/ferminews (accessed 4 February 1999). I was in attendance at the symposium described in this article and heard a similar narrative repeated by a number of speakers, including Norman Ramsey. The open-housing story has also been reiterated to me in several interviews with long-term FNAL employees.

59. Several texts have been written about the site selection process, each offering a slightly different perspective on the issues involved. See Catherine Westfall, "The First 'Truly National Laboratory'"; Anton J. Jachim, *Science Policy Making in the United States*; M. Stanley Livingston, *Early History of the 200 GeV Accelerator*; Theodore Lowi and Benjamin Ginsberg, *Poliscide*.

60. Hugh Davis Graham, "The Surprising Career of Federal Fair Housing Law," 215–32.

61. As quoted in Mike Perricone, "Eye Witness to History," www.fnal.gov/pub/ferminews (accessed 4 February 1999).

62. Robert Wilson and Ned Goldwasser, "Human Rights Policy."

63. Lillian Hoddeson and Catherine Westfall, "Thinking Small in Big Science," 457–94.

64. "Chronological Timeline of Fermilab Accomplishments," FNAL Web site, www.fnal.gov.

65. "Chronological Timeline of Fermilab Accomplishments," FNAL web site, www.fnal.gov.

66. Hoddeson and Westfall, 457–94, 467.

67. Ibid., 471.

68. "Beam or Bust: Tales from the Early Day," presentation given at the Symposium Celebrating Ned Goldwasser's Eightieth Birthday, 10 March 1999.

69. "Chronological Timeline of Fermilab Accomplishments," FNAL Web site, www.fnal.gov.

70. "Historical-Timeline," http://www.fnal.gov/pub/about/whatis/timeline.html (accessed 12 January 2006).

71. In one compelling episode on 4 June 1983, immediately prior to an important meeting that would decide the fate of future particle accelerators in the U.S., the *New York Times* editorial page blasted the American physics community for losing the race with Europe and called on George Keyworth, White House science adviser and head of the Office of Science and Technology Policy, to rally the "team" and advocate for new facilities and more physics research funding.

72. It is important to note that the funds the community had to work with did not compare in real terms to the funding of labs following World War II. While the years leading up to the SSC gave physicists reason to be optimistic, actual funding levels had been flat for a number of years when measured in constant dollars and adjusted for inflation. Many in the physics community argue that the need to tie increases in funding to construction of new facilities on undeveloped sites is the most significant problem in funding physics research today.

73. Judy Jackson, interviewed by author, 6 June 1997.

74. Adrienne Kolb, interviewed by author, 6 June 1997.

75. The final act in the drama of the SCC took place 19 October 1993. See "Kill the Superconducting Super Collider," *Congressional Record*.

76. At the time of the publication of this book, Fermilab is no longer the world's most powerful accelerator. It lost this distinction when the Large Hadron Collider at CERN began operating in September 2008. The unrelenting quest for power is motivated by a desire to find the Higgs boson, a critical explanatory element of the standard model of physics. Shortly after the cancellation of the SSC and the discovery of the top quark, there was some mild controversy about the energy range in which the Higgs might be located. The unexpected mass of the top quark led some physicists to suggest that even the LHC may not be sufficiently powerful to produce the Higgs boson (See Jay Orear, "Does Top Mass Rule Out Higgs at LHC," 15). However, wider opinion seemed to favor a "light Higgs" that would likely be produced by the LHC and perhaps even by the upgraded but less powerful main injector at Fermilab (reports to Fermilab users and the High Energy Physics Advisory Panel, 1997 Fermilab Users Meeting, at Fermilab National Accelerator Laboratory, Batavia, 14–17 July 1997).

77. Even though the SSC was to be funded separately, cost overruns eventually did threaten proposed scheduled upgrades for existing laboratories. Fermilab narrowly avoided congressional cutbacks that eventually would have ended its viability as a research facility. Political pressure from the Illinois congressional delegation ultimately secured the future of the "main injector," an upgrade that should allow scientists to run fixed target and accelerator experiments simultaneously and thus effectively double the "scientific output" currently possible at Fermilab.

78. "Chronological Timeline."

79. "Mission Statement."

80. Fermilab's new main injector allows researchers to conduct collider and fixed-target experiments simultaneously. The current fixed-target work, however, must be conducted at much lower energies than in previous eras, thus dramatically altering the nature of the research questions that can be asked and answered in these experiments.

81. The time needed for the completion and commissioning of the main injector and recycler has allowed for upgrades to other facilities as well, most notably the two collider detectors, CDF and D-Zero.

82. Simply put, "commissioning" refers to the process of putting the accelerator through its paces in preparation for research use. Of particular importance is the delivery of beam at consistent, usable energies for the detectors recording particle collisions.

83. Neal Lane, presentation given at the 1997 Fermilab Users Meeting, Fermi National Accelerator Laboratory, Batavia, 14 July 1997.

Chapter Two: Mapping the Boundaries of the Laboratory

1. See J. Ziman, *Public Knowledge*; P. M. Rattansi, "The Social Interpretation of Science in the Seventeenth Century," 1–33; Rupert Hall, "Science, Technology and

Utopia in the Seventeenth Century," 33–53; John A. Campbell, "Scientific Revolution and the Grammar of Culture," 351–76.

2. I use the term "science studies" not to denote any formal program of research, but rather as an umbrella term referring to the variety of analyses on the subject of science. This group includes, but is not limited to, work in sociology, history, philosophy, rhetoric, critical, and feminist studies.

3. See the work generated from the National Science Foundation's Public Understanding of Science project; "Public Understanding" and "Public Attitudes" sections contained in the biennial *Science and Engineering Indicators* reports sponsored by the National Science Board; and the work of Jon Miller, director of the International Center for the Advancement of Scientific Literacy. A few outside summaries of these studies also exist. See Georgine Pion and Mark Lipsey, "Public Attitidues towards Science and Technology," 303–16.

4. According to the most recent NSF data, 45 percent of Americans describe themselves as interested in science and technology. This number stands in contrast, however, to a steady decline in the importance of science and technology in relationship to other topics in the opinion of those surveyed. For a detailed account, see chapter 7, "Science and Technology: Public Attitudes and Understanding," *Science and Engineering Indicators 2004,* http://www.nsf.gov/statistics/seind04/pdf/c07.pdf (accessed 18 January 2006).

5. Measures of attentiveness have changed over the years. Most recently it has been assessed in terms of *inclination*—participants report themselves to be very interested in these issues; *knowledge*—participants are considered "very well informed"; and *behavior*—participants regularly expose themselves to information about science and technology (see Miller). The evolution of this measure, while not unexpected in a longitudinal study, raises concerns about the generalizability of the findings reviewed here.

6. Pion and Lipsey, 215.

7. Ibid., 215.

8. Thomas Goodnight's work on the public sphere provides valuable insight into the functions of the technical sphere in contemporary society. As Goodnight points out, while some decision making is and should be subjected to technical evaluation, many issues are relegated to the technical sphere as a means to bypass the public sphere.

9. See, for instance, Karl Popper, *The Logic of Scientific Discovery.*

10. This caricature is largely sociological.

11. In his text on demarcation rhetorics, *Defining Science,* Taylor calls for such an extension of the research in this area: "A logical extension of the framework elaborated here would be a context in which consensually-accepted lines of external and internal demarcation must be redrawn" (224).

12. See Taylor, *Defining Science,* 3–20 and 222–29.

13. Ibid., 135–74.

14. Anne Holmquest, "The Rhetorical Strategy of Boundary Work," 235–58.

15. Peter Galison, *The Disunity of Science*.
16. See Thomas Gieryn, "Boundary Work and the Demarcation of Science," 402–20.
17. Robert Merton, *The Sociology of Science;* Michael Mulkay, "Norms and Ideology in Science," 535–52; Michael Mulkay, *Science and the Sociology of Knowledge*.
18. Thomas Gieryn, "Boundary Work and the Demarcation of Science," 781–795. Further development of this idea can be found in Holmquest, "The Rhetorical Strategy of Boundary Work," 1999 and Taylor, *Defining Science*, 1991.
19. Thomas Gieryn, "Boundary Work and the Demarcation of Science," 781–95.
20. Ibid., 792.
21. Ibid., 792.
22. Holmquest, "The Rhetorical Strategy of Boundary Work," 235–58.
23. Ibid., 252.
24. Ibid., 253.
25. Taylor, "Scientific Community," 402–20.
26. Michael Polanyi, as quoted in Taylor, *Defining Science*, 133.
27. Ibid., 133.
28. Ibid., *227*.
29. Ibid., 16.
30. The term "visible scientist" was coined by Rae Goodell, *The Visible Scientists*. This study explores the social capital and behavior of prominent members of the scientific community.
31. Michael Mulkay, "The Mediating Role of the Scientific Elite," 445–70.
32. Thomas M. Lessl, "The Priestly Voice," 195. See also Thomas M. Lessl, "Science and the Sacred Cosmos," 175–87.
33. Panofsky has testified before Congress on this issue on numerous occasions. Prior to my first interview with him in the summer of 2002, when he was eighty-three years old, he had just returned from giving testimony on the Comprehensive Test Ban Treaty.
34. Taylor, *Defining Science*, 129.
35. Lessl, "Priestly Voice," 183–97.
36. Ibid., 129.
37. Robert Hariman, "Decorum, Power, and the Courtly Style," 149–72.
38. Carole Blair, Marsha Jeppeson, and Enrico Pucci, "Public Memorializing in Postmodernity," 282.
39. Ibid., 283.
40. Catherine Kohler Reissman, *Narrative Analysis*, 8–15.
41. Quoted in James A. Holstein and Jaber F. Gubrium, "Phenomenology, Ethnomethodology, and Interpretive Practice," 263.
42. See Holstein and Gubrium, "Phenomenology, Ethnomethodology, and Interpretive Practice," 267.
43. Hans George Gadamer, *Philosophical Hermeneutics*, 62.

44. Bruno Latour's *Science in Action* is one of the best examples of the application of observational techniques for the study of science in society. Latour's work provides a critical insider's perspective on the everyday work of scientists, including the discursive practices that shape scientific inquiry and change.

45. James A. Holstein and Jaber F. Gubrium, *The Active Interview*, 16–18.

46. Thomas Lindlof, *Qualitative Communication Research Methods*; Holstien and Gubrium, *Active Interview*, 16–18.

47. Ian Hodder, "The Interpretation of Documents and Material Culture," 394.

48. As detailed in the text, this study utilized interviewing techniques in concert with rhetorical analysis to discern how experience, discourse, and interpretation function to construct and stabilize Fermilab, its work, and social conceptions of basic scientific research. These approaches are combined as a means to acknowledge and explain the interplay between language and experience and thus explore directly the experiences that prompt support, opposition, or ambivalence to science as well as attend to the discourse that gives structure to such experiences. It is hoped that this integrated design might provide critical insight into both the evolving relationship between science and society and more specific instances of scientific rhetoric.

Interviewed participants for the initial portion of the Fermilab study were identified from among the two primary categories of nonscientist: visitors and employees. Observation was employed to the extent that it was useful as a means to understand better the discursive environment at Fermilab, the culture of high-energy physics, and the unique experiences of both visitors and employees.

Approximately thirty participants were chosen from among those taking self-tours of the facility. The self-tour consists of three films, several static exhibits, and a walking tour of the fifteenth floor. The researcher observed participants from a distance but did not accompany visitors on their tour. Following the self-tour, participants were interviewed individually or in groups to assess their impressions of the tour, the facility, and the research being done there. Interviews were approximately twenty to thirty minutes in length.

Thirty-one employees of Fermilab were interviewed in order to develop a detailed understanding of the work being done at the facility and the organization's attempts to communicate that work to the public. Particular attention was paid to the employees' assessment of the past and future of Fermilab and high-energy physics and the public's role in that history. These interviews did not exceed ninety minutes in length.

Textual analysis of interview transcriptions was conducted in order to identify the stocks of knowledge visitors and employees draw on to make sense of their time at Fermilab. The analytical framework employed for the purposes of this analysis was derived from the data: particular attention was given to observed points of contact and disparity among the experience and interpretation of those working at Fermilab and those visiting the facility. Research at Fermilab was approved by the University of Georgia's Institutional Review Board. Subsequent interviews with employees at Lawrence Berkeley National Laboratory, Brookhaven National Laboratory, Jefferson National Laboratory, and Stanford Linear Accelerator Center may be referenced in this study. These interviews were done with the approval of the University of Iowa Institutional Review Board.

Chapter Three: Rhetoric, Persuasion, and the Sublime Laboratory

1. Robert Hariman, "Terrible Beauty and Mundane Detail," 10–18.
2. Ibid.
3. David Nye, *American Technological Sublime*, xiii.
4. Theories of rhetoric have periodically explored the persuasive impact of the sublime. Most scholars cite Longinus's *On the Sublime*, probably written in the second century C.E., as one of the earliest attempts to identify, understand, and theorize the peculiar emotional response that we recognize as the sublime. In his comments on Longinus's work, D. A. Russell offers the following description of a sublime text: "It makes us proud and joyful, as though we had ourselves created it. It stands the test of repeated reading. . . . It is irresistible and memorable." (Longinus, *On the Sublime*). Such strong emotional response reveals the persuasive capacity of the sublime. As James Aune described it in his work on Lincoln's use of the American sublime, "Longinus' notion of the Sublime seems to capture that sense of release from self, that *ek-stasis*, or ecstasy, when one is taken out of everyday life" (Aune, "Lincoln and the American Sublime," 14–19). This sense of another world or plane of experience is central to our understanding of the sublime. The persuasive capacity of rhetoric increases proportionally to the degree that it can transport the audience to another space or perspective from which they might view the world around them. Longinus's concept of the sublime was strongly linked to the beautiful. The eighteenth-century British rhetorician and member of parliament Edmund Burke, however, sought to distinguish the sublime and the beautiful, theorizing what he believed to be related but distinct emotional reactions. Noting how fear functions to produce sublime response, Burke provided an interpretation of the sublime that still resonates today: "Whatever is fitted in any sort to excite the ideas of pain, and danger, that is to say, whatever is in any sort terrible, or is conversant about terrible objects, or operates in a manner analogous to terror, is a source of the sublime; that is, it is productive of the strongest emotion which the human mind is capable of feeling. . . . When danger or pain press too nearly, they are incapable of giving any delight, and are simply terrible; but at certain distances, and with certain modifications, they may be, and they are delightful, as we every day experience" (Edmund Burke, *A Philosophical Enquiry*, section 7). Burke retained Longinus's focus on extreme emotion, but he theorized how even terror works to produce sublime effect if we feel safe. In doing so, Burke suggested the defining characteristic of the sublime was *duality*. Burke's analysis also emphasized the various manifestations of the sublime—vastness, magnitude, difficulty—and theorized how those qualities in an object, or attributed to one, elicited the described emotional effect. Like Longinus, Burke recognized the persuasive capacity of the sublime. However he framed his analysis in terms of social control rather than the art of persuasion. He was interested like many contemporary rhetorical scholars in the political effectivity of the sublimity of kingship and political authority generally. Whereas Burke focused on the

relationship between external forces, qualities, or objects and the emotional response they elicited in bodies, Kant responded by focusing on the sublime as a manifestation of the mind and its human capacity for reason. Kant produced a simpler and more general taxonomy of sublime experience, focusing on two categories, the mathematical and the dynamic sublime. The mathematical sublime was characterized by "boundlessness," the excess of scale or scope one might experience when contemplating a vast mountain range. The dynamic sublime was defined by excess power. Nuclear explosions or rocket launches can be considered examples of the dynamic sublime, if only in the technological mode. In offering these categories, Kant's focus was not on the objects that produced the sublime, but rather on our internal response. In his quest to understand the sublime as a function of reason, he offered the following explanation:

> Sublimity, therefore, does not reside in anything of nature, but only in our mind, in so far as we can become conscious that we are superior to nature within, and therefore also to nature without us.... Everything that excites this feeling in us, e.g. the *might* of nature which calls forth our forces, is called then (although improperly) the sublime. Only by supposing this idea in ourselves and in reference to it are we capable of attaining to the idea of sublimity of that Being which produces respect in us, not merely by the might that it displays in nature, but rather by means of the faculty which resides in us of judging it fearlessly and of regarding our destination as sublime in respect to it (Immanuel Kant, *Observations on the Beautiful and Sublime*).

For Kant our rationality resides in our understanding of the relationship between ourselves and the sublime object. Whereas Longinus located the persuasive power of the sublime in its capacity to take us outside of ourselves and our everyday existence, Kant saw the rational resolution of the sublime—the opportunity to go within ourselves and discover our own rationality—as the ultimate reward. While offering sometimes radically different interpretations of the sublime, each of these theorists locates sublime experience in the opposition of ordinary and extraordinary states of being. The same variety of interpretation exists today but more often in relation to the technological sublime than the natural.

5. David Nye, *American Technological Sublime*, xiii.
6. Ibid., xviii.
7. Ibid., xviii.
8. Ibid., 4.
9. Ibid., 4.
10. Sheldon Glashow, as quoted in "Race for the Top," *Nova* (PBS video, 1988).
11. It should be noted that alchemic metaphors, images, and discourses have a long history. The impact of this rhetorical frame has been studied as it relates to the development of nuclear weapons but somewhat surprisingly has not been extensively explored in relation to other areas of physics. For additional insight into the image of the alchemist in science fiction, see Spencer R. Weart, *Nuclear Fear*.

12. *Fermilab: Discovering the Nature of Nature.*
13. Mircea Eliade, *The Forge and the Crucible,* 8–9. These themes have also been analyzed with respect to nuclear power and nuclear weapons. For an interesting discussion of the sublime in these contexts, see Weart, *Nuclear Fear.*
14. Eliade, *The Forge and the Crucible,* 8.
15. *Fermilab: Boundless Horizons,* 10.
16. Ibid., 10.
17. Ibid., 48.
18. *Pursuit of the Fundamental / Welcome To Fermilab.*
19. Ibid., 8.
20. John Huth as quoted in *Fermilab: Boundless Horizons,* 26.
21. Peter Berger, *The Sacred Canopy.*
22. Fermilab is part of the Department of Energy's system of National Environmental Resource Parks. The program conducts studies of the environmental systems at various DOE labs. Given a history of contamination, the research at most sites focuses on how natural systems rebound from environmental catastrophe. Fermilab's site serves as a control in that it is the only NERP that until recently has not suffered from any known contamination. Visitors are introduced to this concept of the NERP through a video entitled *The Fragile Balance.*
23. Rich Orr, in *Ferminews,* 2 July 1999.
24. Ibid.
25. Despite its independent appearance, the laboratory was supported by a consortium of universities from the start. Credit for the consortium idea goes to Leon Lederman.
26. Robert R. Wilson, *Starting Fermilab,* 17–18.
27. Personal communication with Nancy Kerrigan, 10 March 1999.
28. B. Grypton, "Fermilab: Where Science Is Art," 38.
29. Ibid.
30. FNAL press release #88–719–8.
31. B. Grypton, "Fermilab: Where Science Is Art," 44.
32. Ibid., 38.
33. Wilson, *Starting Fermilab,* 14.
34. During a visit to the laboratory in August 1997, Peter Rosen referred to Wilson's influence by suggesting possible alternative names for Wilson Hall. Among his lighthearted suggestions were "Minster Wilson," "The Cathedral of Saint Robert," and "The Cross of Batavia." On a more serious note he acknowledged that the building "symbolized the power and presence of science in the lives of the people."
35. Completely enclosed, this dark, paneled room seems a bit out of place in the light filled atmosphere of Wilson Hall. However "warm" the wood paneling may be in comparison to the poured concrete that dominates the building, it is the conference room and not Wilson Hall that is repeatedly singled out for criticism. Reflecting on its cold, corporate character, one laboratory employee likened the room to "a funeral home."

36. Leon M. Lederman, "Revolution in Science Education," 44.

37. Leon M. Lederman, *The God Particle*, 14–15.

Chapter Four: Practice and Perception

1. The exploits on both sides of the "science wars" provide ample evidence of the uptake of science studies discourse. My experience suggests that while few scientists may actually read primary texts in science studies, many are aware of the discourse. The responses I have encountered vary from amused admissions that they are in fact quite a bit like the descriptions offered up in many sociological studies to biting critiques such as those exemplified in *Dreams of a Final Theory* by Steven Weinberg.

2. NUMI is the acronym for "Neutrinos in the Main Injector." The NUMI project is tangential to the main facility, sending a beam of neutrinos under Wisconsin to a deserted mine shaft in Minnesota. Such long-baseline experiments allow researchers to detect and analyze the neutrino oscillation.

3. Catherine Westfall, "Collaborating Together," 169, 177

4. Sharon Traveek, *Beamtimes and Lifetimes*; Peter Louis Galison, *Image and Logic*.

5. Participant no. 24, Fermi National Accelerator Laboratory employee. Interview by author.

6. Ibid.

7. Participant no. 14, Fermi National Accelerator Laboratory employee. Interview by author.

8. Ibid.

9. It has been reported that Lederman himself used "TNL" (Truly National Laboratory) as a not-so-subtle stab at BNL (Brookhaven National Laboratory) and its exclusion of researchers from outside the BNL community.

10. Participant no. 19, Fermi National Accelerator Laboratory employee. Interview by author.

11. Participant no. 14, Fermi National Accelerator Laboratory employee. Interview by author.

12. Participant no. 19, Fermi National Accelerator Laboratory employee. Interview by author.

13. Participant no. 20, Fermi National Accelerator Laboratory employee. Interview by author, 30 September 1997. Tape recording. Fermi National Accelerator Laboratory, Batavia.

14. The term "cowboy culture" has been used repeatedly by those at Fermilab and also by employees at other laboratories. It activates a number of symbolic references from the laboratory's frontier rhetoric to the horse-related hobbies of both Wilson and Lederman. Most significant, it describes "how things used to be" when researchers could venture off alone without the oversight of Occupational Health and Safety personnel or practical restrictions of a large collaboration.

15. It is important to note that this was a rather retrograde notion of "evolution."

16. The failure to develop an international coalition to fund the SSC was itself an interesting rhetorical episode. The need to secure significant contributions from foreign countries such as Japan could not be met within the limitations set in place by

other discourses operative in the 1980s and early 90s, most notably the need to frame the SSC as an American achievement or a vehicle for job creation and economic development in the United States.

17. Participant no. 19, Fermi National Accelerator Laboratory employee. Interview by author.

18. Participant no. 25, Fermi National Accelerator Laboratory employee. Interview by author.

19. Participant no. 15, Fermi National Accelerator Laboratory employee. Interview by author.

20. Participant no. 12, Fermi National Accelerator Laboratory employee. Interview by author.

21. Participant no. 15, Fermi National Accelerator Laboratory employee. Interview by author.

22. Participant no. 20, Fermi National Accelerator Laboratory employee. Interview by author.

23. Participant no. 24, Fermi National Accelerator Laboratory employee. Interview by author.

24. Participant no. 20, Fermi National Accelerator Laboratory employee. Interview by author.

25. Participant no. 21, Fermi National Accelerator Laboratory employee. Interview by author.

26. Participant no. 24, Fermi National Accelerator Laboratory employee. Interview by author.

27. Participant no. 15, Fermi National Accelerator Laboratory employee. Interview by author.

28. Participant no. 22, Fermi National Accelerator Laboratory employee. Interview by author.

29. Participant no. 26, Fermi National Accelerator Laboratory employee. Interview by author.

30. These were terms repeatedly used by interview participants to characterize the high energy physics community's experience of the SSC.

31. Participant no. 22, Fermi National Accelerator Laboratory employee. Interview by author.

32. Participant no. 14, Fermi National Accelerator Laboratory employee. Interview by author.

33. Participant no. 18, Fermi National Accelerator Laboratory employee. Interview by author.

34. Participant no. 22, Fermi National Accelerator Laboratory employee. Interview by author.

35. Participant no. 25, Fermi National Accelerator Laboratory employee. Interview by author.

36. Participant no. 26, Fermi National Accelerator Laboratory employee. Interview by author.

37. Participant no. 14, Fermi National Accelerator Laboratory employee. Interview by author.
38. Participant no. 26, Fermi National Accelerator Laboratory employee. Interview by author.
39. Participant no. 12, Fermi National Accelerator Laboratory employee. Interview by author.
40. Participant no. 19, Fermi National Accelerator Laboratory employee. Interview by author.
41. Participant no. 12, Fermi National Accelerator Laboratory employee. Interview by author.
42. Participant no. 14, Fermi National Accelerator Laboratory employee. Interview by author.
43. Participant no. 16, Fermi National Accelerator Laboratory employee. Interview by author.
44. Participant no. 26, Fermi National Accelerator Laboratory employee. Interview by author.
45. Participant no. 18, Fermi National Accelerator Laboratory employee. Interview by author.
46. Participant no. 12, Fermi National Accelerator Laboratory employee. Interview by author.
47. Ibid.
48. Participant no. 16, Fermi National Accelerator Laboratory employee. Interview by author.
49. Participant no. 14, Fermi National Accelerator Laboratory employee. Interview by author.
50. Participant no. 15, Fermi National Accelerator Laboratory employee. Interview by author.
51. Participant no. 24, Fermi National Accelerator Laboratory employee. Interview by author.
52. Participant no. 25, Fermi National Accelerator Laboratory employee. Interview by author.
53. Participant no. 18, Fermi National Accelerator Laboratory employee. Interview by author.
54. Participant no. 19, Fermi National Accelerator Laboratory employee. Interview by author.
55. Participant no. 19, Fermi National Accelerator Laboratory employee. Interview by author.
56. Participant no. 21, Fermi National Accelerator Laboratory employee. Interview by author.
57. Participant no. 24, Fermi National Accelerator Laboratory employee. Interview by author.
58. Participant no. 14, Fermi National Accelerator Laboratory employee. Interview by author.

59. Participant no. 15, Fermi National Accelerator Laboratory employee. Interview by author.
60. Participant no. 14, Fermi National Accelerator Laboratory employee. Interview by author.
61. Participant no. 19, Fermi National Accelerator Laboratory employee. Interview by author.
62. Participant no. 20, Fermi National Accelerator Laboratory employee. Interview by author.
63. Ibid.
64. Participant no. 13, Fermi National Accelerator Laboratory employee. Interview by author.
65. Participant no. 25, Fermi National Accelerator Laboratory employee. Interview by author.
66. Participant no. 21, Fermi National Accelerator Laboratory employee. Interview by author.
67. Participant no. 18, Fermi National Accelerator Laboratory employee. Interview by author.
68. Ibid.
69. Participant no. 16, Fermi National Accelerator Laboratory employee. Interview by author.
70. Participant no. 21, Fermi National Accelerator Laboratory employee. Interview by author.
71. Participant no. 16, Fermi National Accelerator Laboratory employee. Interview by author.
72. Participant no. 20, Fermi National Accelerator Laboratory employee. Interview by author.
73. Participant no. 20, Fermi National Accelerator Laboratory employee. Interview by author.
74. Participant no. 13, Fermi National Accelerator Laboratory employee. Interview by author.
75. Ibid.
76. Participant no. 19, Fermi National Accelerator Laboratory employee. Interview by author.
77. Participant no. 20, Fermi National Accelerator Laboratory employee. Interview by author.
78. Comments of Rich Orr as reprinted in *Ferminews*, 2 July 1999.
79. Participant no. 12, Fermi National Accelerator Laboratory employee. Interview by author.
80. Participant no. 13, Fermi National Accelerator Laboratory employee. Interview by author.
81. Participant no. 25, Fermi National Accelerator Laboratory employee. Interview by author.
82. Ibid.

83. Lederman, *The God Particle.*
84. Participant no. 16, Fermi National Accelerator Laboratory employee. Interview by author.
85. Participant no. 24, Fermi National Accelerator Laboratory employee. Interview by author.
86. Participant no. 12, Fermi National Accelerator Laboratory employee. Interview by author.
87. Participant no. 27, Fermi National Accelerator Laboratory employee. Interview by author.
88. Participant no. 18, Fermi National Accelerator Laboratory employee. Interview by author.

Chapter Five: Stakeholders, Self-Tours, and Communication after the SSC

1. While the direct appeals to the general public were new phenomena, the arguments in favor of limiting scientific funding were well known. Drawing on the long history of particle physics' relationship to the government (detailed in chapter 2), those who spoke out against the SSC cited lack of funds, lack of oversight, and more pressing and relevant social concerns as their primary reasons for opposition.
2. Participant no. 27, Fermi National Accelerator Laboratory employee. Interview by author.
3. Ibid.
4. While this book focuses on nonscientist visitors, it is important to note that the boundary enacts this relationship with respect both to scientists from other disciplines and to lay people. As the boundary exists for the purposes of distinguishing the FNAL community from a variety of competing interests, the right to transgress the boundary is only granted to those who espouse or support the world view of those "inside." The self-tour can thus serve as a test for scientists and nonscientists alike.
5. For example, information about the discovery of the top quark is treated as both a mark of identity and an example of the type of work done at the lab and thus is located in both areas.
6. Mission statement as printed on fig. 10.
7. A fourth film, *The Fragile Balance,* will not be treated here. This video, produced by the DOE, detailed the DOE's National Environmental Resource Park program, in which FNAL participates. Most visitors observed and interviewed for this study did not watch this video in its entirety.
8. *Pursuit of the Fundamental / Welcome to Fermilab,* produced and directed by Visual Media Services, Fermi National Accelerator Laboratory.
9. Ibid.
10. Ibid.
11. Ibid.
12. Ibid.
13. Ibid.

14. In using the word "single," I am referring to the sought-after particle (the top quark, the Higgs boson, and so on). In each collision, of course, there are always many particles created.

15. *Pursuit of the Fundamental / Welcome to Fermilab.*

16. Ibid.

17. Ibid.

18. Research began with formal interviews of visitors to Fermilab's public areas, particularly the self-tour area on the fifteenth floor of Wilson Hall. Interviews were conducted from September 1997 through January 1998. Fourteen male and sixteen female visitors, ranging in age from twenty to seventy-seven, participated. Among those interviewed, fourteen reported having college degrees, six reported having some college education, and three had no college education. All were high school graduates. Of those reporting college degrees or education, fifteen had taken course work in science and/or engineering. Most participants cited physics and astronomy, natural sciences, or biology as their primary scientific interest. Subsequent observational and historical-archival research was conducted at FNAL and other national laboratories from 1998 to 2003.

19. This may have been the last FNAL open house of its kind, where some ten thousand visitors passed through Wilson Hall and toured many of the laboratories experimental facilities. Post-9/11 security measures make it almost impossible to admit visitors who do not have business at the laboratory or are not escorted by a laboratory employee. These new security measures have also meant the end of the self-tour. Public visits to the laboratory are now restricted to one day a month when neighbors are invited to tour the fifteenth floor and hear a lecture on physics research.

20. Participants V18, 19, and 20, self-tour visitors. Interview by author.

21. Participants V26 and 27, self-tour visitors. Interview by author.

22. The second response was clearly shaped by a question about "social function"; however, the first discussion and others like it emerged without prompting during the course of the interview. The majority of participants talked about science in terms of technological products or useful, applicable knowledge.

23. Participants V11 and 12, self-tour visitors. Interview by author.

24. Ibid.

25. Participant V6, self-tour visitor. Interview by author.

26. Ibid.

27. Ibid.

28. Ibid.

29. Participant V5, self-tour visitor. Interview by author.

30. Ibid.

31. Ibid.

32. Ibid.

33. Participant V10, self-tour visitor. Interview by author.

34. Participant V18, 19, 20, self-tour visitors. Interview by author.

35. Participants V16 and 17, self-tour visitors. Interview by author. The laboratory did install an extensive exhibit on the history of nuclear medicine that emphasized the role of high-energy physics and particle accelerators in the development of medical technologies and treatments. As the rhetorical history of the SSC demonstrates, however, nuclear medicine is a contested discourse. While the foundations for many technologies can rightfully be traced to developments in accelerator technology, the work of design and application has often taken place in other fields and thus any rhetoric must be carefully crafted to respect subdisciplinary boundaries within physics when making arguments about nuclear medicine. Such boundaries are largely meaningless for visitors, however.

36. Participants V21 and 22, self-tour visitors. Interview by author.
37. Participants V16 and 17, self-tour visitors. Interview by author.
38. Participant V1, self-tour visitors. Interview by author.
39. Participant V7, self-tour visitor. Interview by author.
40. Participants V26 and 27, self-tour visitors. Interview by author.
41. Participant V22, self-tour visitor. Interview by author. A similar point is made by V21.
42. Participants V18 and 19, self-tour visitors. Interview by author.
43. Participant V6, self-tour visitor. Interview by author.
44. Participant V12, self-tour visitor. Interview by author.
45. Participant V10, self-tour visitor. Interview by author.

Chapter Six: Mapping the Boundaries and Charting a Future

1. Nye, *American Technological Sublime*, 227, 234.
2. Ibid., 234, 235.
3. Ibid., 255.
4. In this regard, the relation between the war on terror and the marketplace, including the issue of nuclear proliferation, is a topic that one hopes will be taken up by future scholars.

Bibliography

Published Work

Aune, James Arnt. "Lincoln and the American Sublime," *Communication Reports* 1 (Winter 1988), 14–19.

Berger, Peter. *The Sacred Canopy*. New York: Anchor, 1967.

Blair, Carole, Marsha Jeppeson, and Enrico Pucci. "Public Memorializing in Postmodernity: The Vietnam Veterans Memorial as Prototype," *Quarterly Journal of Speech* 77 (August 1991): 263–88.

Burke, Edmund. *A Philosophical Enquiry into the Origin of Our Ideas of the Sublime and the Beautiful*. Oxford: Oxford University Press, 1990.

Butler, Sharon. "Path to the Energy Frontier? A Very Large Hadron Collider," *Ferminews*, 12 September 1997, 1.

Campbell, John Angus. "Scientific Revolution and the Grammar of Culture: The Case of Darwin's *Origin*." *Quarterly Journal of Speech* 72 (November 1986): 351–76.

Chaffee, C. David. "Can Big Science Claim Credit?" *Science* 253 (September 1991): 1204.

Cozzens, Susan, and Thomas Geiryn, eds. *Theories of Science in Society*. Bloomington: Indiana University Press, 1990.

Deutch, John. "A Super Collision of Interests," *Technology Review* 95 (November 1992): 66–77.

Eliade, Mircea. *The Forge and the Crucible*. Chicago: University of Chicago Press, 1978.

———. *The Sacred and the Profane: The Nature of Religion*. New York: Harcourt, Brace, 1959.

Fermilab: Boundless Horizons. Washington, D.C.: University Research Associates, 1990.

Fermilab: Discovering the Nature of Nature. Batavia: Fermi National Accelerator Laboratory, 1994.

Gadamer, Hans George. *Philosophical Hermeneutics*. Berkeley: University of California Press, 1977.

Galison, Peter. *The Disunity of Science: Boundaries Contexts and Power*. Stanford: Stanford University Press, 1996.

———. *Image and Logic*. Chicago: University of Chicago Press, 1997.

Galison, Peter, and Bruce Hevley, eds. *Big Science: The Growth of Large Scale Research*. Stanford: Stanford University Press, 1992.

Gieryn, Thomas. "Boundary Work and the Demarcation of Science from Non-Science: Strains and Interests in Professional Ideologies of Scientists," *American Sociological Review* 48 (October 1983): 781–95.

Goodell, Rae. *The Visible Scientists*. Boston: Little, Brown, 1975.

Graham, Hugh Davis. "The Surprising Career of Federal Fair Housing Law," *Journal of Policy History* 12, no. 2 (2000): 215–32.

Greenberg, Daniel. *The Politics of Pure Science*. New York: Plume, 1967.

Grypton, B. "Fermilab: Where Science Is Art," *Science Digest* 94 (February 1986): 38.

Hall, Rupert A. "Science, Technology and Utopia in the Seventeenth Century." In *Science and Society: 1600–1900*, ed. Peter Mathias, 33–53. Cambridge: Cambridge University Press, 1972.

Hariman, Robert. "Decorum, Power, and the Courtly Style," *Quarterly Journal of Speech* 78 (May 1992): 149–72.

———. "Terrible Beauty and Mundane Detail: Aesthetic Knowledge in the Practice of Everyday Life," *Argumentation and Advocacy* 35 (Summer 1998): 10–18.

Hepburn, R.W. "Sublime." In *The Oxford Companion to Philosophy*, ed. Ted Honderich. Oxford: Oxford University Press, 1995, 857–58.

Hodder, Ian. "The Interpretation of Documents and Material Culture." In *The Handbook of Qualitative Research*, eds. Norman Denzin and Yvonna Lincoln, 393–402. Thousand Oaks: Sage, 1994.

Hoddeson, Lillian, and Adrienne Kolb. "The New Frontier in the Chicago Suburbs: Settling Fermilab," *Illinois Historical Journal* 88, no. 1 (1995): 2–18.

Hoddeson, Lillian, and Catherine Westfall. "Thinking Small in Big Science: The Founding of Fermilab, 1960–1972." *Technology and Culture* 37, no. 3 (1996): 457–94.

Holmquest, Anne. "The Rhetorical Strategy of Boundary Work." *Argumentation* 4 (August 1990): 235–58.

Holstein, James A., and Jaber F. Gubrium. "Phenomenology, Ethnomethodology, and Interpretive Practice." In *The Handbook of Qualitative Research*, eds. Norman Denzin and Yvonna Lincoln, 262–72. Thousand Oaks: Sage, 1994.

———. *The Active Interview*. Thousand Oaks: Sage, 1995.

Jachim, Anton J. *Science Policy Making in the United States and the Batavia Accelerator*. Carbondale: Southern Illinois University Press, 1968.

Jackson, Judy. "DOE Removes Brookhaven Contractor," *Ferminews*, 16 May 1997, 8–9.

———. "Reaching for the Muon," *Ferminews*, 1 November 1996, 1.

Kant, Immanuel. *Observations on the Beautiful and Sublime*. Berkeley: University of California Press, 1960.

Kevles, Daniel J. *The Physicists: The History of a Scientific Community in Modern America*. Cambridge: Harvard University Press, 1995.

"Kill the Superconducting Super Collider." Congressional Record.—House, H8102, 1993 (accessed 12 January 2006).

Latour, Bruno. *Science in Action*. Cambridge: Harvard University Press, 1987.

Lederman, Leon. *The God Particle*. New York: Delta, 1993.

———. "Open Letter to Colleagues Who Publicly Opposed the SSC." *Physics Today* 47 (March 1994): 9–11.

———. "Revolution in Science Education: Put Physics First," *Physics Today* 54 (September 2001): 44.
Leslie, Stuart W. *The Cold War and American Science: The Military-Industrial-Academic Complex at MIT and Stanford*. New York: Columbia University Press, 1993.
Lessl, Thomas M. "Heresy, Orthodoxy, and the Politics of Science." *Quarterly Journal of Speech* 74 (February 1988): 18–34.
———. "The Priestly Voice." *Quarterly Journal of Speech* 75 (1989): 195.
———. "Science and the Sacred Cosmos: The Ideological Rhetoric of Carl Sagan." *Quarterly Journal of Speech* 71 (1985): 175–87
Lindlof, Thomas R. *Qualitative Communication Research Methods*. Thousand Oaks: Sage, 1995.
Livingston, M. Stanley. *Early History of the 200 GeV Accelerator*. Batavia: University Research Associates, 1968.
Longinus. *On the Sublime*, commentary by D. A. Russell. Oxford: Oxford University Press, 1964.
Lowi, Theodore, and Benjamin Ginsberg. *Poliscide*. New York: Macmillan, 1976.
Malamud, Ernest, and Michael Albrow. "Exploring Nature with the Very Large Hadron Collider: The Next Step beyond the LHC." http://www-ap.fnal.gov/VLHC/vlhcpub/pubs1-100/1/popular.html
Merton, Robert. *The Sociology of Science: Theoretical and Empirical Investigations*. Chicago: University of Chicago Press, 1973.
Midgley, Mary. *Science as Salvation*. London: Routledge, 1992.
Miller, Jon D. *The American People and Science Policy*. Elmsford: New York, 1983.
Mulkay, Michael. "The Mediating Role of the Scientific Elite." *Social Studies of Science* 6 (May 1976): 445–70.
———. "Norms and Ideology in Science." *Social Science Information* 15 (1975): 535–52
———. *Science and the Sociology of Knowledge*. London: Allen and Unwin, 1979.
National Science Board. *Science and Engineering Indicators—1982*. Washington D.C.: GPO, 1983.
———. *Science and Engineering Indicators—1984*. Washington D.C.: GPO, 1985.
———. *Science and Engineering Indicators—1986*. Washington D.C.: GPO, 1987.
———. *Science and Engineering Indicators—1988*. Washington D.C.: GPO, 1989.
———. *Science and Engineering Indicators—1990*. Washington D.C.: GPO, 1991.
———. *Science and Engineering Indicators—1992*. Washington D.C.: GPO, 1993.
———. *Science and Engineering Indicators—1994*. Washington D.C: GPO, 1995.
———. *Science and Engineering Indicators—1996*. Washington D.C: GPO, 1997.
Nye, David. *American Technological Sublime*. Cambridge: MIT, 1996.
Orear, Jay. "Does Top Mass Rule Out Higgs at LHC?" *Physics Today* 48 (August 1995): 15.
Peterson, Ivars. "Shattering the SSC." *Science* 144 (October 1993): 276.
Pion, Georgine, and Mark Lipsey. "Public Attitudes towards Science and Technology: What Have the Surveys Told Us?" *Public Opinion Quarterly* 45 (Fall 1981): 303–16.

Popper, Karl. *The Logic of Scientific Discovery*. London: Hutchinson, 1968.
Price, Don K. *The Scientific Estate*. Cambridge: Harvard University Press, 1965.
Quigg, Chris. "A Little Bit of the Gods." Presentation given in March 1994 at Fermilab at the International Symposium and Tribute in Honor of Robert R. Wilson on His Eightieth Birthday. http://lutece.fnal.gov/Notes/SplashGods.pdf (accessed 3 September 2008).
Rattansi, P. M. "The Social Interpretation of Science in the Seventeenth Century." In *Science and Society: 1600–1900*, ed. Peter Mathias, 1–33. Cambridge: Cambridge University Press, 1972.
Reissman, Catherine Kohler. *Narrative Analysis*. Newbury Park: Sage, 1993.
Sears, John. *Sacred Places: American Tourist Attractions in the Nineteenth Century*. New York: Oxford University Press, 1989.
Self Tour Brochure. N.p.: Department of Public Affairs, Fermi National Accelerator Laboratory, n.d.
Seidel, Robert W. "The Origins of the Lawrence Berkeley Laboratory." In *Big Science*, Peter Galison and Bruce Hevley, eds, 21–45. Stanford: Stanford University Press, 1992.
Taylor, Charles Alan. *Defining Science: A Rhetoric of Demarcation*, Madison: University of Wisconson Press, 1996.
———. "Defining the Scientific Community: A Rhetorical Perspective on Communication." *Communication Monographs* 58 (1991): 402–20.
Traveek, Sharon. *Beamtimes and Lifetimes: The World of High Energy Physicists*. Cambridge: Harvard University Press, 1988.
Weart, Spencer R. *Nuclear Fear* (Cambridge: Harvard University Press, 1988).
Weinberg, Alvin. "The Impact of Large-Scale Science in the United States." *Science* 134 (July 14 1961): 161–64.
Westfall, Catherine. "Collaborating Together: The Stories of TPC, UA1, CDF, and CLAS." *Historical Studies in the Physical and Biological Sciences*, 32, no. 1 (2001): 169, 177.
———. "The First 'Truly National Laboratory': The Birth of Fermilab." Ph.D. diss., Michigan State University, 1988.
Wilson, Robert R. *The Humanness of Physics*. Batavia, Ill.: FNAL, 1978.
———. *Starting Fermilab*. Batavia: Fermi National Accelerator Laboratory, 1997.
Wilson, Robert, and Ned Goldwasser. "Human Rights Policy." 1968; reprinted in *Fermilab: Boundless Horizons*. Washington D.C.: University Research Associates, 1990.
Ziman, J. *Public Knowledge: An Essay Concerning the Social Dimension of Science*. Cambridge: Cambridge University Press, 1968.

Films

Pursuit of the Fundamental / Welcome To Fermilab, produced and directed by Visual Media Services, Fermi National Accelerator Laboratory.
The Fragile Balance, produced and directed by Visual Media Services, Fermi National Accelerator Laboratory.

Interviews, Presentations, and Other Unpublished Work

"FNAL Press Release no. 88-719-8." Public Information Office, Fermi National Accelerator Laboratory.

Goldwasser, Ned. "Reminiscences." Presentation given at the Symposium Celebrating Ned Goldwasser's 80th Birthday, 10 March 1999.

Jackson, Judy. Public affairs director, Fermi National Accelerator Laboratory. Interviewed by author. Tape recording, 6 June 1997. Fermi National Accelerator Laboratory, Batavia.

Kolb, Adrienne. Laboratory archivist, Fermi National Accelerator Laboratory. Interviewed by author. Tape recording, 6 June 1997. Fermi National Accelerator Laboratory, Batavia.

Lane, Neal. Presentation given at the 1997 Fermilab Users Meeting, Fermi National Accelerator Laboratory, Fermilab National Accelerator Laboratory, Batavia, Illinois, 14 July 1997.

Orr, Rich. "Beam or Bust: Tales from the Early Days." Presentation given at the Symposium Celebrating Ned Goldwasser's 80th Birthday, Fermi National Accelerator Laboratory, Batavia, Illinois, 10 March 1999.

Participant no. 1. Fermi National Accelerator Laboratory employee. Interview by author, 21 July 1997. Tape recording. Fermi National Accelerator Laboratory, Batavia.

Participant no. 2. Fermi National Accelerator Laboratory employee. Interview by author, 24 July 1997. Tape recording. Fermi National Accelerator Laboratory, Batavia.

Participant no. 3. Fermi National Accelerator Laboratory employee. Interview by author, 24 July 1997. Tape recording. Fermi National Accelerator Laboratory, Batavia.

Participant no. 4. Fermi National Accelerator Laboratory employee. Interview by author, 24 July 1997. Tape recording. Fermi National Accelerator Laboratory, Batavia.

Participant no. 7. Fermi National Accelerator Laboratory employee. Interview by author, 12 August 1997. Tape recording. Fermi National Accelerator Laboratory, Batavia.

Participant no. 8. Fermi National Accelerator Laboratory employee. Interview by author, 21 August 1997. Tape recording. Fermi National Accelerator Laboratory, Batavia.

Participant no. 9. Fermi National Accelerator Laboratory employee. Interview by author, 21 August 1997. Tape recording. Fermi National Accelerator Laboratory, Batavia.

Participant no. 10. Fermi National Accelerator Laboratory employee. Interview by author, 21 August 1997. Tape recording. Fermi National Accelerator Laboratory, Batavia.

Participant no. 11. Fermi National Accelerator Laboratory employee. Interview by author, 26 August 1997. Tape recording. Fermi National Accelerator Laboratory, Batavia.

Participant no. 11a. Fermi National Accelerator Laboratory employee. Interview by author, 26 August 1997. Tape recording. Fermi National Accelerator Laboratory, Batavia.

Participant no. 12. Fermi National Accelerator Laboratory employee. Interview by author, 26 August 1997. Tape recording. Fermi National Accelerator Laboratory, Batavia.

Participant no. 13. Fermi National Accelerator Laboratory employee. Interview by author, 28 August 1997. Tape recording. Fermi National Accelerator Laboratory, Batavia.

Participant no. 14. Fermi National Accelerator Laboratory employee. Interview by author, 11 September 1997. Tape recording. Fermi National Accelerator Laboratory, Batavia.

Participant no. 15. Fermi National Accelerator Laboratory employee. Interview by author, 11 September 1997. Tape recording. Fermi National Accelerator Laboratory, Batavia.

Participant no. 16. Fermi National Accelerator Laboratory employee. Interview by author, 11 September 1997. Tape recording. Fermi National Accelerator Laboratory, Batavia.

Participant no. 17. Fermi National Accelerator Laboratory employee. Interview by author, 11 September 1997. Tape recording. Fermi National Accelerator Laboratory, Batavia.

Participant no. 18. Fermi National Accelerator Laboratory employee. Interview by author, 18 September 1997. Tape recording. Fermi National Accelerator Laboratory, Batavia.

Participant no. 19. Fermi National Accelerator Laboratory employee. Interview by author, 25 September 1997. Tape recording. Fermi National Accelerator Laboratory, Batavia.

Participant no. 20. Fermi National Accelerator Laboratory employee. Interview by author, 30 September 1997. Tape recording. Fermi National Accelerator Laboratory, Batavia.

Participant no. 21. Fermi National Accelerator Laboratory employee. Interview by author, 30 September 1997. Tape recording. Fermi National Accelerator Laboratory, Batavia.

Participant no. 22. Fermi National Accelerator Laboratory employee. Interview by author, 30 September 1997. Tape recording. Fermi National Accelerator Laboratory, Batavia.

Participant no. 23. Fermi National Accelerator Laboratory employee. Interview by author, 2 October 1997. Tape recording. Fermi National Accelerator Laboratory, Batavia.

Participant no. 24. Fermi National Accelerator Laboratory employee. Interview by author, 2 October 1997. Tape recording. Fermi National Accelerator Laboratory, Batavia.

Participant no. 25. Fermi National Accelerator Laboratory employee. Interview by author, 2 October 1997. Tape recording. Fermi National Accelerator Laboratory, Batavia.

Participant no. 26. Fermi National Accelerator Laboratory employee. Interview by author, 7 October 1997. Tape recording. Fermi National Accelerator Laboratory, Batavia.

Participant no. 27. Fermi National Accelerator Laboratory employee. Interview by author, 7 October 1997. Tape recording. Fermi National Accelerator Laboratory, Batavia.

Participant no. 28. Fermi National Accelerator Laboratory employee. Interview by author, 9 October 1997. Tape recording. Fermi National Accelerator Laboratory, Batavia.

Participant no. 29. Fermi National Accelerator Laboratory employee. Interview by author, 9 October 1997. Tape recording. Fermi National Accelerator Laboratory, Batavia.

Participant no. 30. Fermi National Accelerator Laboratory employee. Interview by author, 14 October 1997. Tape recording. Fermi National Accelerator Laboratory, Batavia.

Participant no. 31. Fermi National Accelerator Laboratory employee. Interview by author, 14 October 1997. Tape recording. Fermi National Accelerator Laboratory, Batavia.

Participant 0V1. Open house visitor. Interview by author, 13 September 1997. Tape recording. Fermi National Accelerator Laboratory, Batavia.

Participant 0V2. Open house visitor. Interview by author, 13 September 1997. Tape recording. Fermi National Accelerator Laboratory, Batavia.

Participant 0V3. Open house visitor. Interview by author, 13 September 1997. Tape recording. Fermi National Accelerator Laboratory, Batavia.

Participant V1. Self-tour visitor. Interview by author, 7 August 1997. Tape recording. Fermi National Accelerator Laboratory, Batavia.

Participant V2. Self-tour visitor. Interview by author, 7 August 1997. Tape recording. Fermi National Accelerator Laboratory, Batavia.

Participant V3. Self-tour visitor. Interview by author, 7 August 1997. Tape recording. Fermi National Accelerator Laboratory, Batavia.

Participant V4. Self-tour visitor. Interview by author, 14 September 1997. Tape recording. Fermi National Accelerator Laboratory, Batavia.

Participant V5. Self-tour visitor. Interview by author, 14 September 1997. Tape recording. Fermi National Accelerator Laboratory, Batavia.

Participant V6. Self-tour visitor. Interview by author, 1 November 1997. Tape recording. Fermi National Accelerator Laboratory, Batavia.

Participant V7. Self-tour visitor. Interview by author, 7 November 1997. Tape recording. Fermi National Accelerator Laboratory, Batavia.

Participant V8. Self-tour visitor. Interview by author, 8 November 1997. Tape recording. Fermi National Accelerator Laboratory, Batavia.

Participant V9. Self-tour visitor. Interview by author, 8 November 1997. Tape recording. Fermi National Accelerator Laboratory, Batavia.

Participant V10. Self-tour visitor. Interview by author, 8 November 1997. Tape recording. Fermi National Accelerator Laboratory, Batavia.
Participant V11. Self-tour visitor. Interview by author, 8 November 1997. Tape recording. Fermi National Accelerator Laboratory, Batavia.
Participant V12. Self-tour visitor. Interview by author, 8 November 1997. Tape recording. Fermi National Accelerator Laboratory, Batavia.
Participant V13. Self-tour visitor. Interview by author, 8 November 1997. Tape recording. Fermi National Accelerator Laboratory, Batavia.
Participant V14. Self-tour visitor. Interview by author, 8 November 1997. Tape recording. Fermi National Accelerator Laboratory, Batavia.
Participant V15. Self-tour visitor. Interview by author, 8 November 1997. Tape recording. Fermi National Accelerator Laboratory, Batavia.
Participant V16. Self-tour visitor. Interview by author, 8 November 1997. Tape recording. Fermi National Accelerator Laboratory, Batavia.
Participant V17. Self-tour visitor. Interview by author, 8 November 1997. Tape recording. Fermi National Accelerator Laboratory, Batavia.
Participant V18. Self-tour visitor. Interview by author, 8 November 1997. Tape recording. Fermi National Accelerator Laboratory, Batavia.
Participant V19. Self-tour visitor. Interview by author, 8 November 1997. Tape recording. Fermi National Accelerator Laboratory, Batavia.
Participant V20. Self-tour visitor. Interview by author, 8 November 1997. Tape recording. Fermi National Accelerator Laboratory, Batavia.
Participant V21. Self-tour visitor. Interview by author, 9 November 1997. Tape recording. Fermi National Accelerator Laboratory, Batavia.
Participant V22. Self-tour visitor. Interview by author, 9 November 1997. Tape recording. Fermi National Accelerator Laboratory, Batavia.
Participant V23. Self-tour visitor. Interview by author, 9 November 1997. Tape recording. Fermi National Accelerator Laboratory, Batavia.
Participant V24. Self-tour visitor. Interview by author, 9 November 1997. Tape recording. Fermi National Accelerator Laboratory, Batavia.
Participant V25. Self-tour visitor. Interview by author, 9 November 1997. Tape recording. Fermi National Accelerator Laboratory, Batavia.
Participant V26. Self-tour visitor. Interview by author, 9 November 1997. Tape recording. Fermi National Accelerator Laboratory, Batavia.
Participant V27. Self-tour visitor. Interview by author, 9 November 1997. Tape recording. Fermi National Accelerator Laboratory, Batavia.
Reports to Fermilab Users Group and the High Energy Physics Advisory Panel, 1997 Fermilab Users Meeting, Fermi National Accelerator Laboratory, Fermilab National Accelerator Laboratory, Batavia, Illinois, 14–17 July 1997.

Index

aesthetics, xiii, 26, 68, 162. *See also* sublime; technological sublime; Fermi National Accelerator Laboratory, rhetorics of
American Medical Association (AMA), 109
antimatter, 59
antiprotons, 4, 11, 29, 59, 60, 78, 117, 132, 135, 167n3. *See also* proton-antiproton particle collisions
Apollo Space Program, 19
Argonne National Laboratory, 17
argument, spheres of, 39, 172n5. *See also* Goodnight, G. Thomas
Associated Universities, Inc. (AUI), 17
atomic bomb, 58, 65, 138–39, 161
Atomic Energy Commission (AEC): Brookhaven National Laboratory and, 17–18, 26; commissioning of FNAL by, 10, 12, 15, 17, 26; Lawrence Berkeley National Laboratory and, 26; NAL site declaration of and, 24; NAL design and, 26; National Science Foundation (NSF) and, 16; Ramsey Report and, 20; Wilson, Robert, and, 66–67, 169n53
"Atoms for Peace," 19
Aune, James, 175n4

Barrington, Illinois, 24
Batavia, Illinois, 1, 28–29, 147, 171nn76, 83, 177n34
Berkeley Radiation Laboratory (Rad Lab). *See* Lawrence Berkeley National Laboratory
BeV (billion electron volt) machines, 18, 22, 77–78
"big science." *See* science, "big" and "small"
Black Panthers, 25
Blair, Carole, 49, 173n38
BNL. *See* Brookhaven National Laboratory
bottom quark. *See* quark, bottom
boundaries of science, defining. *See* "boundary work"
"boundary work" (boundaries of science, defining), 43, 46–49, 84, 96, 119, 134, 137, 146, 156, 159, 172n14, 173nn16–22
Braidwood plants, 162
Brookhaven National Laboratory (BNL), 17–18, 23–24, 45, 50, 103–4, 124, 167n1, 169n35, 174n48, 178n9
Bulletin of the Atomic Scientists, 15, 21
Burke, Edmund, 56–57, 175n4
Bush, George W., 45, 105
Bush, Vannevar, 12–17, 77, 80, 153, 168. See also *Science: The Endless Frontier*

Campbell, John A., 172n1
CDF. *See* collision detector facilities
CERN. See *Conseil Européen pour la recherche nucléaire*
Cockroft-Walton machine, 117, 134

cold war: FNAL and, 20–21, 28, 58, 110; postwar scientific developments and, 19–20; research versus weapons laboratories during, 58, 86, 110
"collidercentrism," 89; SSC and, 27–28
collision detector facilities (CDFs), 78, 82, 85, 92, 132, 135, 171n81. *See also* D0; E-786
Comprehensive Test Ban Treaty, 173n33
Conseil Européen pour la Recherche Nucléaire (CERN), 11, 27, 29, 79, 112, 124, 171n76
contamination, environmental, at national laboratories. *See* national laboratories, environmental contamination of
Cornell University, 66

D0, 78, 82, 85. *See also* collision detector facilities
Defining Science, 41–42, 172n11–12, 173nn18, 26, 34. *See also* Charles Alan Taylor
Denver, Colorado, 24
Department of Energy, 28–31, 51, 65, 91–92, 100, 103–5, 118, 150
Dilbert (comic strip), 148
Dirksen, Sen. Everett, 24–25
Dr. Strangelove (film), 163
DOE. *See* Department of Energy
"doomsday stories" about high-energy physics, 86–87

E-786. *See* collision detector facilities
Einstein, Albert, 80
Eisenhower, Dwight D., 19–20, 44; "Atoms for Peace" plan and, 19–20
Eliade, Mircea, 60–61, 177n13
engineering: civil, 17; genetic, 157; science and, 120, 112
Enlightenment, the 21
Environmental Protection Agency (EPA), 160
European Organization for Nuclear Research. *See* Conseil Européen pour la Recherche Nucléaire (CERN)

Fair Housing Act, 25
Faust, Faustian, 7–8, 66, 145, 163
Federation of American Scientists, 21
Fermi, Enrico, 4, 15, 27
Fermilab. *See* FNAL
Feynman, Richard, 4
Feynman Computing Center, 132
Fermi National Accelerator Laboratory (FNAL): authorization of, 26; budget of, 11; cold war and, 19–21, 27–28-2 58, 62, 86, 110–11, 238, 169n38; comparative analysis of, with other national laboratories, 166; commissioning of, 10; construction, costs of, 15; consumer choice/corporate profit, as recent theme in, 149–50, 163; "cowboy culture" as early theme in, 81, 90, 178n14; distinctiveness of among other national laboratories, 153–54; ethos and, 33, 48, 64, 77, 90, 93, 104, 153, 157; experimenter ethic of, 80; Fair Housing Act and, 25; films about, 6, 123, 132–36, 174n48; frontier themes in, 3, 24, 27, 30–31, 70, 72, 153, 178n14; grounds/physical space of, 47, 62 "open houses," held at, 99, 106–8, 114, 137, 183n19; "inner" versus "outer space," distinction between 124, 126, 136; "insider" versus "outsider," 6, 32, 36–40, 85; monologic, 152, 164; mystification, as, 58, 127, 163–65; narrative quality of, 3; nationalist versus globalist, as theme in, 110, 113, 125; "priestly voice" and, 43–46, 147, 149, 173n32, n35; property values of surrounding area, 103; rhetorics of, aesthetic dimension of, xiii, 3, 26, 48–50, 54–55, 66, 68, 70, 71, 74, 130, 133, 154, 156, 162; sculpture and (at FNAL), 47, 49, 62, 68 70, 77, 132; sublimation

(alchemical), as theme of, 3; sublime, as theme of, 55–56, 59–65, 68, 70, 73, 77, 79, 85, 122, 131, 135–37, 142, 148–50, 154, 162, 164; utopian quality of, 156. *See also* interviews; Gonzalez, Angela; self-tour
Foucault pendulum, 73
frontier and mythology/imagery of and science, 3, 12–13, 17, 22, 24, 31, 34, 67, 153, 168n11, 178n14; and "energy," 10, 27, 2–30, 78–79, 110, 111, 125, 126. *See also* Bush, Vannevar; *Science: The Endless Frontier*

Gadamer, Hans George, 173n43
Galileo, 102
Galison, Peter, 38, 86, 168nn10, 13, 173n15, 178n4, 185
GeV (billion electron volts), 27, 30, 78, 170n59
Gieryn, Thomas, 38–41, 173nn16, 18, 19
Ginsberg, Benjamin, 170n59
Glashow, Sheldon, 58, 176n10
God Particle, The, 81, 114, 168n15, 178n37, 182n83. *See also* Lederman, Leon
Goldwasser, Ned, 24–26, 169n54, 170n62, n68
Gonzalez, Angela, 3–4, 7–8, 48, 72–76, 127, 167n2
Goodell, Rae, 43–44, 173n30
Goodnight, G. Thomas, 39, 172n8. *See also* arguments, spheres of
Government Performance and Results Act, 30
Graham, Hugh Davis, 170n60
Greenberg, Daniel, 19, 169n41
gross national product (GNP), 113
Grypton, B., 177n28, n31
Gubrium, Jaber, 173nn41, 42, 174nn45, 46

hadronic colliders, 5, 11, 29, 88, 92, 110, 112, 171n76
Hariman, Robert, 48, 55, 173n37, 175n1
Hevley, Bruce, 168n13
Higgs boson, 171n76, 183n14
Hodder, Ian, 174n47
Hoddeson, Lillian, 26, 168nn10–11, 170nn63, 66
Holmquest, Anne, 38–41, 172n14, 173n18, n22
Holstein, James, 173nn41–42, 174n45
Huth, John, 61, 177n20

Illinois Math and Science Academy, 80
International Center for Advancement of Scientific Literacy, 172n3
International style (architecture), 71
interviews: of employees of Fermi National Accelerator Laboratory, 3, 31, 85–103, 114–16; 168n10, 170n58; of visitors to Fermi National Accelerator Laboratory, 1–3, 31, 123, 136–37, 141–46, 149–50, 156–59; method of conducting, 47, 50 52–53, 174n48

Jachim, Anton J., 170n59
Jefferson National Laboratory, 174n48
Jeppesen, Marsha, 49, 173n38
Johnson, Lyndon B.: and authorization of FNAL, 37; and commissioning of FNAL, 10; and dedication of Stanford Linear Accelerator, 21; May-Johnson Bill, 14–15; and open housing laws, 24–25

Kant, Immanuel, 56–57, 161, 176. *See also* sublime
Kennedy, Robert, 25
Kevles, Daniel, 13–16, 18–20, 168nn10, 16, 25, 169nn36, 42, 47, 49. See also *The Physicists*
Kerrigan, Nancy, 177n27
Keyworth, George, 170n71
Kilgore, Harley, 13–16
King, Jr., Martin Luther, 24–25
Kolb, Adrienne, 168n10, 171n74
Krige, John, 168n10

laboratories, national. *See* National Laboratory System
Laboratory Services Department, 120
Lane, Neal, 31, 171n83
large hadron collider (LHC). *See* hadronic colliders
Latour, Bruno, 173n44
Lawrence, Ernest O., 15, 17, 23, 26, 65, 165nn33–34, 165n55, 174n48. *See also* Lawrence Berkeley National Laboratory
Lawrence, John, 168n34
Lawrence Berkeley National Laboratory (LBL), 17–18, 23, 26, 29, 34, 65–66, 168nn33–34, 169n55, 174n48
Lawrence Berkeley Radiation Laboratory (Rad Lab). *See* Lawrence Berkeley National Laboratory
Lawrence Livermore National Laboratory, 168n33, 169n55
Lederman, Leon, 10, 23, 27–28, 45, 55, 64–65, 77–82, 114, 119, 156, 168n15, 177n25, 178nn9, 14, 36, 37, 182n83. *See also God Particle, The*
Leonardo da Vinci, 8
leptons, 1, 60, 93, 133
Leslie, Stuart, 169n38
Lessl, Thomas, xiv, 44, 45, 173nn32–33, 35
Lindlof, Thomas, 174n46
LHC. *See* hadronic colliders
Limon, Peter, 4
Lincoln, Abraham, 175n4
linear accelerators, 19, 28, 44, 67, 124, 134, 167nn1–2, 174n48. *See also* Stanford Linear Accelerator (SLAC)
Lipsey, Mark, 172nn3, 6
Livermore. *See* Lawrence Livermore National Laboratory
Livingston, M. Stanley, 170n59
Lofgren, Edward, 23, 26, 66
Longinus, 56, 175–76, n4
Los Alamos National Laboratory, 14, 17, 44, 57, 65–66, 72, 167n1
Lowi, Theodore, 170n59

Magnuson, Warren, and National Science Foundation Bill, 16
main injector, 28–30, 85, 125, 132, 140, 171nn76, 77, 80, 81, 178n2
main ring, 27, 30, 85, 134
Malamud, Ernie, 121
Manhattan Project, 44, 58, 138, 167n1
Marburger, John, 44–45
May-Johnson Bill, 14–15
McMahon, Brien, 15
Merton, Robert, 173n17
mesons, 18
MeV (million electron volts), 28
Midgley, Mary, 12
Midwestern Universities Research Association (MURA) 18, 22–23
military-industrial complex, 19–20, 22, 36, 66–68, 97, 104–6, 138, 153–54, 162
Miller, Jon, 172, n3
Morton Aircraft Corporation, 119
Mulkay, Michael, 44, 173nn17, 31
MURA. *See* Midwestern Universities Research Association

National Accelerator Laboratory, (NAL). *See* Fermi National Accelerator Laboratory
National Environmental Resource Parks (NERP), 177n22, 182n7
National Laboratory System, 11, 17, 119, 167n1, 168n10, 169n55; aesthetics of, 156; commissioned by Atomic Energy Commission, 17; contractor management of, 18, 148; costs of, 15; crisis of rationality in, 156; environmental contamination and, 43, 105, 119, 160–65, 177n22; history of, 11, 17–21; imagining, 2–4; marketplace rhetorics and, 163–64; national interest and, 24–30; public and, 118–19, 155, 159, 160; military-industrial complex and, 57, 161, 164; multipurpose, 57; science and technology, as blurred in 24; SSC and, 171n77; trust in, public, 104–5;

war on terror and, 161; weapons work, as inherently implicated in, 161–62
National Science Board (NSB), 33–34, 172n3
National Science Foundation (NSF), 12, 14–16, 31, 172n3
neutrinos 11, 178n2. *See also* Neutrinos in the Main Injector Project (NUMI)
Newman, James R., 15
New Republic, 13
New York Times, 170n71
"next machine," rhetoric of, 28–29, 79, 92, 96, 105, 111, 125, 153
NSB. *See* National Science Board
NSF. *See* National Science Foundation
nuclear age, 34
nuclear annihilation, 162
nuclear explosions, 176
nuclear materials, extraordinary (sublime) nature of, 62, 66, 161
nuclear forces, 18
nuclear medicine, 17, 184n35
nuclear physics 15, 18, 20, 161
nuclear power, 19, 34, 104–5, 161, 177n13
nuclear proliferation, 19, 184n4; specter of, 163
nuclear submarines, 104
nuclear technologies, 44, 57. 62, 161
nuclear weapons, 19, 45, 104, 138, 161, 176n11, 177n13
Neutrinos in the Main Injector (NUMI) project, 11, 178n2
Nye, David, 55–57, 161, 175n3, 176nn5–9, 184n1

Oak Ridge National Laboratory, 17, 57, 167n1
Occupational Health and Safety Administration (OSHA), 105, 178n14
Oddone, Pier, 30
Office of Scientific Research and Development, 12
Office of War and Mobilization, 15
Oppenheimer, Robert, 15, 44–45
Orr, Richard, 27, 65, 112, 177n23, 181n78
OSHA. *See* Occupational Health and Safety Administration

P-Bars. *See* antiprotons
Panofsky, Wolfgang, 19, 44–45, 173n33
particle accelerators, 1, 17, 29, 56, 68, 79, 103, 112, 117, 134, 170n71, 184n35
Pauling, Linus, 43
People, John, 28, 30
Perricone, Mike, 170n58, n61
Physicists, The, 13, 168nn16–23, 25–33, 36–37, 42–43, 47–49. *See also* David Kevles
physics: and accelerator, 66–67; astro-, 134; applied, 112; basic, 16–17, 19, 57, 65, 118; cultural status of, 20 62; education in, 106–7; and engineering, 17; epistemic value of, 6–7; establishment as, 15, 77–78, 80, 118; funding of, 10, 22, 56, 66, 97, 119; future of, 8, 15, 21, 31, 44, 88, 118; high-energy, xiii, 2–3, 6, 10–11, 19, 28–32, 40–41, 45, 48, 51, 55–61, 66–67, 77–82, 84–88, 91, 93, 95–97, 101, 105–6, 112–15, 122, 124, 127, 133, 134; history of, 5, 15, 35–36, 82, 114; humanism and, 6–8, 67, 115.; international collaboration in, 110; military-industrial complex and, 33, 66, 68; neutrino, 92; nuclear, 18; particle, 66–67, 78, 93–94, 97, 108, 113, 115, 120, 122, 124, 133, 134; plasma, 134; publics and, 43, 58, 101–2; rhetoric of, 4, 57, 84, 125, 135; unified theory in, 60; wartime and, 11–12, 27; weapons and, 57, 86–87, 104; World Wide Web and, 112. *See also* military industrial complex; nuclear weapons; Wilson, Robert
Pion, George, 35, 172nn3, 6
Polanyi, Michael, 173n26
Popper, Karl, 172n9
Price, Donald, 15–16, 20, 21, 168nn10, 24, 169nn45, 48, 50. *See also Scientific Estate, The*

proton-antiproton particle collisions, 29, 78
public relation, 106–9, 113, 120
public: and FNAL, xiii, 2–11, 14, 20, 30–31, 36, 42–44, 46–48, 55–58, 96–99, 101, 115, 118, 136, 155; interest, 15–18, 21–22, 63, 97, 101, 154; policy, 31; and project of National Science Foundation, 172n3; relations, 106–9, 113, 120; sphere, 35–36, 38–40, 43, 46; stakeholders as, 36, 42, 82, 119, 149, 152, 163, 166; and understanding of science, 5, 14, 33, 35, 38, 98, 102, 106, 108, 137, 146, 154, 157, 159–60
Pucci, Enrico, 49, 173n38
Pugwash conference, 110

quantum mechanics, 58, 123
quarks, 1, 60, 93, 133, 143, 145, 150; bottom, 27, 77–78; top, 21, 29–30, 114, 127–29, 171n76, 182n5, 183n14
Quigg, Chris, 4, 167n3

research and development (R and D) facilities, 12, 20, 78, 145
Rabi, I. I., 17
Rad Lab. *See* Lawrence Berkeley National Laboratory
Ramsey committee, 22
Ramsey, Norman, 17, 24, 77, 170n58. *See also* Ramsey report
Ramsey report, 20–22, 32
Rattansi, P. M., 171n1
reactors (nuclear), 19–20, 104; research on, 17, 20; graphite, 17, 19, 169n35
Reagan, Ronald, 27
Reissman, Catherine Kohler, 173n40
rhetoric as style and narrative, 46–50
rhetorical situation, 3, 93, 121, 165–66
rhetorics of FNAL. *See* Fermi National Laboratory, rhetorics of
risk, 103
Roosevelt, Franklin D., 12–13
Rosen, Peter, 177n34

Sagan, Carl, 43–44
"Saturday Morning Physics" program, 80. *See also* Lederman, Leon
Schutz, Alfred, 50
science: "big" versus "small," 3, 11, 20–21, 28, 77, 84, 91, 168n13, 170n63; education and, 107–10; engineering and, 20, 112; gender and, 34; public attitudes toward (as measured by National Science Foundation), 33–35; social construction of, 9. *See also* "boundary work"; Bush, Vannevar; *Science: The Endless Frontier*
science and society, relationship between, 1–2, 21, 33–35, 37
Science: The Endless Frontier, 14, 17, 80, 168n12. *See also* Bush, Vannevar
Scientific American, 98
Scientific Estate, The, 15, 168n24, 169nn45, 48, 50. *See also* Price, Donald
SciTech (museum), 121
sculpture. *See* Fermi National Accelerator Laboratory, sculpture and
Search for Extraterrestrial Intelligence (SETI) program, 44
Seidel, Robert, 17, 168n34
self-tour (of FNAL), 52, 96, 108, 112, 114–15, 117, 121–51, 156–57
September 11, 2001 (9/11), 31, 76, 123, 161, 166, 183n19
sixties (1960s), value system of, 21, 25, 118, 127
SLAC. *See* Stanford Linear Accelerator Center
speed of light, 134
stakeholders, 118–19
standard model (of particle physics) 29, 171n76
Stanford University, 19, 67
sublime: American, 56, 67; Burke's conception of, 175; computers and, 125; definition of, 55–57, 62, 175n4; domestication of, 122, 135, 149, 150; dynamical and mathematical, 124, 130, 132, 176n4; experience of, 98–99,

107, 115, 122–23, 141; experimentalists and, 81; imaginary, and, 57; Kantian conception of, 161, 176n4; knowledge for its own sake and, 142; Longinus's conception of, 175n4; narratives of, 5, 83, 120, 157; natural, 56–58, 61, 63, 136, 150–51, 154; nuclear, 161, 177n13; rhetorics of, 5, 55–56, 64, 114–15, 121–22, 150, 152, 155–56, 161, 163, 175n4; sentimental, 161; sublimation and, 57; technological, 3, 22, 53–64, 67–68, 70–71, 81, 96, 130, 135–37, 141, 151, 154, 163, 176n4; terrible/fearful, and 62, 66, 81, 117, 127, 161; very large and very small in, 135
superconducting supercollider (SSC), 27–29, 79–82, 85, 91, 97, 100, 105, 107, 111–12, 117–121, 123, 125, 127, 153–54, 170n72, 171nn76–77, 178–79n16, 179n30, 182n1, 184n35
synchrotrons, 17–19

Taylor, Charles Alan, 37–38, 40–41, 45–46, 172nn11–12, 173nn18, 25, 26, 34. *See also Defining Science*
technoscience, 4, 8, 117, 133
Tet Offensive (Vietnam War), 25
Tevatron, (TeV), 4, 11, 27–29, 59–60, 70, 78, 117, 124–25, 131, 134
"tiger team" (inspections), 103
top quark. *See* quark, top
Traweek, Sharon, 85, 178n4
"Truly National Laboratory" (TNL), 23, 77, 88, 170n59, 178n9
tritium, 50, 103. *See also* National Laboratory System, environmental contamination of
Truman administration, 14, 16

University of Georgia Institutional Review Board, 174n48
University of Iowa Institutional Review Board, 175n48

URA. *See* Universities Research Associates
Urban League, 25

very large hadronic collider, 112
Vietnam War, 23, 25. *See also* Tet Offensive
Vietnam Veterans War Memorial, 49

W and Z particles, 27
war on terror, 161–62, 164, 184n4
Watkins, Admiral, 104
Waxahachie, Texas, as proposed site of SSC, 28
weapons of mass destruction, 138
Weart, Spencer R., 61, 176n11, 177n13
Weinberg, Alvin, 169n44
Weinberg, Steven, 178n1
Westfall, Catherine, 26, 168n10, 170nn59, 63, 66, 68, 178n3
Weston, Illinois, 24, 26
Wilson, Robert: AEC, testimony before, 169n53; aesthetic sensibility of, 8, 22, 49, 64, 65, 70–71; Angela Gonzalez, and, 3, 48; and design of FNAL, 23, 25, 26, 65; founding director of FNAL, as, 10, 26–27, 49, 64–65, 69–72, 77, 79–80; human rights policy and, 35, 134, 170n62; *Humanness of Physics*, author of, 6; knowledge for its own sake, as advocating, 22, 25; and Lawrence Berkeley National Laboratory (Rad Lab), 17, 26–27, 65; retirement of, 27; rhetorics of FNAL, and 3–5, 8, 23, 47, 55, 59. *See also* rhetorics, of FNAL
Wilson Hall, 54, 62, 73, 121, 123, 125, 177n34, 183nn18, 19, 35. *See also* self-tour
Witherell, Mike, 30
world accelerator movement, 110
wormholes, 98

Z particles. *See* W and Z particles
Ziman, J., 171n1

About the Author

JOANNA S. PLOEGER (1967–2006) was an assistant professor of communication studies at the University of Iowa and California State University, Stanislaus. She served as president of the Association for the Rhetoric of Science and Technology.